Cyber-Physical Systems
AI and COVID-19

Cyber-Physical Systems
AI and COVID-19

Edited by

Ramesh Chandra Poonia
Department of Computer Science, CHRIST (Deemed to be University),
Bangalore, India

Basant Agarwal
Department of Computer Science and Engineering, Indian Institute of
Information Technology (IIIT), Kota, India

Sandeep Kumar
Department of Computer Science and Engineering, CHRIST (Deemed to
be University), Bengaluru, India

Mohammad S. Khan
Network Science and Analysis Lab (NSAL), Department of Computing,
East Tennessee State University, Johnson City, TN, United States

Gonçalo Marques
Polytechnic of Coimbra, ESTGOH, Rua General Santos Costa, Oliveira
do Hospital, Portugal

Janmenjoy Nayak
Aditya Institute of Technology and Management (AITAM) (An
Autonomous Institution), Tekkali, K Kotturu, India

ELSEVIER

ACADEMIC PRESS
An imprint of Elsevier

Academic Press is an imprint of Elsevier
125 London Wall, London EC2Y 5AS, United Kingdom
525 B Street, Suite 1650, San Diego, CA 92101, United States
50 Hampshire Street, 5th Floor, Cambridge, MA 02139, United States
The Boulevard, Langford Lane, Kidlington, Oxford OX5 1GB, United Kingdom

Notices
Knowledge and best practice in this field are constantly changing. As new research and
experience broaden our understanding, changes in research methods, professional practices, or
medical treatment may become necessary.

Practitioners and researchers must always rely on their own experience and knowledge in
evaluating and using any information, methods, compounds, or experiments described herein.
In using such information or methods they should be mindful of their own safety and the safety
of others, including parties for whom they have a professional responsibility.

To the fullest extent of the law, neither the Publisher nor the authors, contributors, or editors,
assume any liability for any injury and/or damage to persons or property as a matter of
products liability, negligence or otherwise, or from any use or operation of any methods,
products, instructions, or ideas contained in the material herein.

British Library Cataloguing-in-Publication Data
A catalogue record for this book is available from the British Library

Library of Congress Cataloging-in-Publication Data
A catalog record for this book is available from the Library of Congress

ISBN: 978-0-12-824557-6

For Information on all Academic Press publications
visit our website at https://www.elsevier.com/books-and-journals

Publisher: Mara Conner
Acquisitions Editor: Sonnini R. Yura
Editorial Project Manager: Mariana L. Kuhl
Production Project Manager: Swapna Srinivasan
Cover Designer: Mark Rogers

Typeset by MPS Limited, Chennai, India

Working together
to grow libraries in
developing countries
www.elsevier.com • www.bookaid.org

Contents

8. A new approach to predict COVID-19 using artificial neural networks 139

Soham Guhathakurata, Sayak Saha, Souvik Kundu, Arpita Chakraborty and Jyoti Sekhar Banerjee

9. Rapid medical guideline systems for COVID-19 using database-centric modeling and validation of cyber-physical systems 161

Mani Padmanabhan

10. Machine learning and security in Cyber Physical Systems 171

Neha V. Sharma, Narendra Singh Yadav and Saurabh Sharma

11. Impact analysis of COVID-19 news headlines on global economy   189

Ananya Malik, Yash Tejas Javeri, Manav Shah and Ramchandra Mangrulkar

12. Impact of COVID-19: a particular focus on Indian education system   207

Pushpa Gothwal, Bosky Dharmendra Sharma, Nandita Chaube and Nadeem Luqman

List of contributors

Basant Agarwal Department of Computer Science and Engineering, Indian Institute of Information Technology (IIIT Kota), Kota, India

Chanchal Ahlawat Department of Computer Science and Engineering, Jaypee Institute of Information and Technology, Noida, India

Amritesh Department of Humanities & Social Sciences, Indian Institute of Technology, Ropar, India

Jyoti Sekhar Banerjee Department of ECE, Bengal Institute of Technology, Kolkata, India

Priyanka Bhaskar Department of Management Studies, Swami Keshvanand Institute of Technology, Management & Gramothan (SKIT), Jaipur, Rajasthan, India

Arpita Chakraborty Department of ECE, Bengal Institute of Technology, Kolkata, India

Avik Chatterjee Department of MCA, Techno India Hooghly, Hooghly, India

Nandita Chaube Gujarat Forensic Sciences University, Gandhinagar, India

Pranjal Chitale D. J. Sanghvi College of Engineering, Mumbai, India

Abhijit Das Department of Information Technology, RCC Institute of Information Technology (RCCIIT), Kolkata, India

Barshan Das Department of MCA, Banaras Hindu University (BHU), Varanasi, India

Basabdatta Das Department of MCA, Techno India Hooghly, Hooghly, India

Jay Gala D. J. Sanghvi College of Engineering, Mumbai, India

Pushpa Gothwal Amity School of Enginnering and Technology, Amity University Rajasthan, Jaipur, India

Soham Guhathakurata Department of CSE, Bengal Institute of Technology, Kolkata, India

Priyanka Harjule Department of Mathematics, Indian Institute of Information Technology (IIIT Kota), Kota, India

Yash Tejas Javeri Computer Engineering Department, Dwarkadas J. Sanghvi College of Engineering, Mumbai University, Mumbai, India

Vijay Jeyakumar Department of Biomedical Engineering, Sri Sivasubramaniya Nadar College of Engineering, Chennai, India

Ruhina Karani D. J. Sanghvi College of Engineering, Mumbai, India

Kaustubh Kekre D. J. Sanghvi College of Engineering, Mumbai, India

Rajalakshmi Krishnamurthi Department of Computer Science and Engineering, Jaypee Institute of Information and Technology, Noida, India

Krishan Kumar Department of Computer Science and Engineering, National Institute of Technology, Srinagar, India

Souvik Kundu Department of Electrical and Computer Engineering, Iowa State University, Ames, IA, United States

Nadeem Luqman Ansal University, New Delhi, India

Ananya Malik Computer Engineering Department, Dwarkadas J. Sanghvi College of Engineering, Mumbai University, Mumbai, India

Ramchandra Mangrulkar Computer Engineering Department, Dwarkadas J. Sanghvi College of Engineering, Mumbai University, Mumbai, India

Shweta Nanda IILM Graduate School of Management, Noida, India

Alok Negi Department of Computer Science and Engineering, National Institute of Technology, Srinagar, India

K. Nirmala Department of Biomedical Engineering, Sri Sivasubramaniya Nadar College of Engineering, Chennai, India

Mani Padmanabhan Faculty of Computer Applications, SSL, Vellore Institute of Technology, Vellore, India

Abhijit S. Pandya Florida Atlantic University, Boca Raton, FL, United States

Riki Patel Florida Atlantic University, Boca Raton, FL, United States

Sunita Rao School of Engineering and Technology, Jaipur National University, Jaipur, Rajasthan, India

Sayak Saha Department of CSE, Bengal Institute of Technology, Kolkata, India

Harshal Sanghvi University School of Sciences, Gujarat University, Ahmedabad, India

Sachin G. Sarate Department of Biomedical Engineering, Sri Sivasubramaniya Nadar College of Engineering, Chennai, India

Sumit Saxena Department of Humanities & Social Sciences, Indian Institute of Technology, Ropar, India

Manav Shah Computer Engineering Department, Dwarkadas J. Sanghvi College of Engineering, Mumbai University, Mumbai, India

Ashish Sharma Department of Computer Science and Engineering, Indian Institute of Information Technology (IIIT Kota), Kota, India

Bosky Dharmendra Sharma Mayoor Private School, Abu Dhabi, United Arab Emirates

Neha V. Sharma Manipal University Jaipur, Jaipur, India

Saurabh Sharma Amity School of Hospitality, Amity University Jaipur, Jaipur, India

Vaishnavi Sharma Department of Chemical Engineering, National Institute of Technology, Raipur, India

Hrishikesh Shenai D. J. Sanghvi College of Engineering, Mumbai, India

Vinita Tiwari Department of Electronics and Communications, Indian Institute of Information Technology (IIIT Kota), Kota, India

Narendra Singh Yadav Manipal University Jaipur, Jaipur, India

Chapter 1

AI-based implementation of decisive technology for prevention and fight with COVID-19

Alok Negi and Krishan Kumar

Department of Computer Science and Engineering, National Institute of Technology, Srinagar, India

1.1 Introduction

Coronaviruses are a wide range of viruses, from the common cold to more extreme diseases such as severe acute respiratory syndrome (SARS-CoV) and middle-eastern respiration syndrome (MERS-CoV) diseases. Many people with pneumonia were admitted from an unexplained cause into Wuhan General Hospital in December 2019. The unknown beta-coronavirus was detected by using objective sequencing in the samples from pneumonia patients. It was a "novel coronavirus" that formed a clade within the subgenus sarbecovirus, the subfamily Ortho-coronavirinae, defined as 2019-nCoV. This novel coronavirus (Guo et al., 2020) is an unidentified entirely new strain in humans and a growing phenomenon from a safety and disease management perspective.

The novelist COVID-19 quickly spread across the world and then became a global epidemic. It has a profound impact on every daily life, global health, as well as the world economy. The new COVID-19 disease is associated with a relatively similar virus family that has SARS and perhaps some kinds of flu or cold. COVID-19 relates to the genera beta-coronavirus depending on all its phylogenetic similarities and genomic properties. Human beta-coronaviruses (SARS-CoV, MERS-CoV, SARS-CoV-2) include several similar characteristics, but also have variations in certain genomic and phenotypic composition, which may affect their pathogenesis. The key symptoms prescribed by the World Health Organisation (WHO) include shortness of breath, fever, cough, and diarrhea nCoV. The upper respiratory signs are

Cyber-Physical Systems. DOI: https://doi.org/10.1016/B978-0-12-824557-6.00008-X

1

less common including runny nose and sneezing. Serious infections may result in pneumonia, failure of the kidneys and death.

Application of the Artificial Intelligence (AI)-driven chest scan has the potential to reduce the increasing burden on radiologists who need to track and evaluate a growing number of chest scans of patients regularly. For the future, the technology will help to predict which patients will most likely require a ventilator or medicine and who can be sent home. Deep learning is a method for studying end-to-end neural networks. It has enabled highly accurate systems for the development of various complex applications (Kumar and Shrimankar, 2017; Kumar, Shrimankar, & Singh, 2018; Kumar, 2019). A convolutional neural network (CNN) developed with various pretrained ImageNet models in Sethy and Behera (2020) for the extraction of high-level features using X-ray images in the chest. Such extracted features were then fed into a Support Vector Machine as a 17machine learning classifier to detect the COVID-19 instances. A COVID-Net based on CNN architecture and transfer learning was introduced in Wang and Wong (2020) for the classification of chest X-ray images among four classes: bacterial, nonCOVID, normal, and 20 COVID-19 viral infections. COVID can best be treated with radiological imaging (Fang et al., 2020) and with imagery (Ai et al., 2020).

To learn the pattern, these systems can understand the dynamic characteristics of input data and self-process such data. Unlike traditional approaches, deep learning allows the data process features and classifications to be extracted simultaneously. Looking at the present problem of coronavirus disease, it has been advised by the WHO, to take preventive measures to safeguard ourselves. One of the main preventive measures taken by the government and WHO is to wear face mask (Leung et al., 2020; MacIntyre and Chughtai, 2020; Chu et al., 2020) while traveling outside along with social distancing. Therefore it has become essential to develop automated applications to find out if anybody wears a mask or not; so that actions can be taken accordingly.

There is no successful vaccine or treatment for the COVID-19 global pandemic and as reported to WHO, there have been 21,026,758 globally confirmed cases with 755,786 deaths as of August 15, 2020. Fig. 1.1 shows the status of COVID-19 as per WHO region (World Health Organization WHO in 2020). By motivating this, we propose two AI-based mechanisms for the prevention of COVID-19. First, a mechanism that identifies those among the crowd, who have not worn a mask. Second, a COVID-19 detection mechanism from the computed tomography (CT) scan images using a deep convolution network that classifies whether a person is a COVID-19 patient or not. The rest part of the chapter is arranged as follows: sections 1.2 and 1.3 describe related work and the proposed work, respectively. Result and analysis are given in section 1.4 followed by conclusion and references.

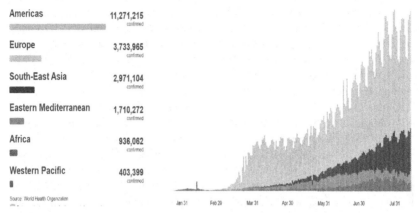

FIGURE 1.1   Global situation by WHO region. *WHO*, World Health Organisation.

1.2   Related work

Han et al. (2020) reported a new effort to recognize COVID-19 through weakly supervised chest CT, some underexamined, but far more feasible, scenario. This research introduced a new, attention-based deep 3D multiinstance learning (AD3D-MIL) for the global pandemic COVID-19 that monitor the poor labels and high generalization. AD3D-MIL provides a deep generator to periodically create deep 3D scenarios, an MIL pooling based on attention that aggregate deep instances into such an insightful bag interpretation, as well as a transformation function to turn the bag interpretation into a joint distribution or Bernoulli distribution for various bag classes.

Fan et al. (2020) proposed the innovative COVID-19 CT scans of the lung infection classification system, dubbed Inf-Net, using implicit reversed attention and clear layer-attention to enhance the detection of infected areas. Besides, the authors provided a semi-supervised viable alternative, Semi-InfNet, to reduce the lack of high reliability labeled data. Experimental findings in the COVID-SemiSeg training dataset and actual CT intensities have shown that the suggested Inf-Net and Semi-Inf-Net features perform better cutting-edge segmentation models as well as order to proceed state-of-the-art efficiency. The suggested model is capable of identifying artifacts with a small level of comparison between pathogens and natural tissues.

Wang et al. (2020) introduced a 3D CT metrics based weakly supervised learning system for the COVID-19 classification and lesion position. For each case, the lung section was partitioned to use a pretrained UNet; then perhaps the segmented 3D lung area was fed into a 3D neural network to determine the risk of infection with COVID-19; the COVID-19 infections are clustered by integrating the triggering regions within the classification network as well as the unsupervised associated components. 499 volumes of CT are being used for preparation and 131 volumes of CT for research. The suggested algorithm provided 0.959 ROC AUC and 0.976 PR AUC. Use a likelihood threshold of 0.5 to distinguish COVID positive and COVID negative, an algorithm achieved a precision of 0.901, a positive predicted value of 0.840 as well as a quite high accuracy of 0.982. The methodology only took 1.93 seconds to analyze the CT volume of a single user using a solely devoted graphics processing unit (GPU).

Chung et al. (2020) reviewed the retrospective case study, chest CT scans of 21 diagnosable patients from China contaminated with 2019 new coronavirus (2019-nCoV) with a premium on recognizing and classifying the most prominent findings. Classic CT findings contained bilateral pulmonary parenchymal glazing and combined pulmonary opacities, often with flattened morphology and peripheral lung coverage. There was no lung cavitation, subtle pulmonary lesions, pleural effusions, no lymphadenopathy. Follow-up examination of a group of patients mostly during the window period of the analysis also indicated a slight to moderate worsening of the condition, as evidenced by a rise of lung opacity and density.

Scientists have shown that wearing face masks helps to reduce COVID-19's spread rate. The wearing of public masks is most effective in stopping viruses spread when enforcement is high. Li et al. (2020) established a methodology that was HGL to overcome the major issue of head pose specification with masks mostly during the COVID-19 endemic problem. This approach uses a study of the color distortion of images as well as a line depiction by retrieving and storing the pixel details from the H-channel mostly in hue, saturation, value (HSV) colored space to differentiate the facial features and mask throughout the image. This enables CNN to learn quite worthwhile information from the input image. The MAFA dataset analysis indicates that the suggested best accuracy approach (front accuracy: 93.64%, side accuracy: 87.17%) was preferable to some other specified methods. The approach to this problem also offers aid in the analysis of multiangle challenges.

Loey, Manogaran, Taha, and Khalifa (2020) proposed a hybrid model using deep and machine learning for the detection of a face mask that has two components. Resnet50 is used for feature extraction as a first component. For the second component, Support Vector Machine, ensemble algorithm, and decision trees are used for the classification purpose. For this work 99.64% testing accuracy is achieved by support vector machine (SVM) in the real-world masked face dataset (RMFD), 99.49% in simulated masked face dataset (SMFD), and 100% in the labeled faces in the wild (LFW) dataset.

Ejaz, Islam, Sifatullah, and Sarker (2019) applied the Principal Component Analysis (PCA) to identify the person in a masked and unmasked face. They found that wearing masks affect the accuracy of face resonation using PCA's extremity. YOLOv3 with Darknet-53 for the face recognition algorithm is introduced by Li, Wang, Li, and Fei (2020). This method has been trained on the CelebA and WIDER FACE datasets of over 600,000 images and received 93.90% accuracy.

1.3 Proposed work

In this study, we have implemented two AI-based preventive measures (face mask detection and CT scanning-based COVID detection) using advanced deep learning models. The first approach is to detect face masks from a crowded zone whereas the other is used to identify COVID-19 patients based on CT scan images using VGG16. Fig. 1.2 shows the overall approach for the proposed work.

Simonyan and Zisserman presented the VGG network architecture in 2014 in their paper (Simonyan and Zisserman, 2014) as shown in Fig. 1.3. This network is distinguished by its straightforwardness, using only 3×3 convolutional layers stacked in rising depth on top of each other. Max pooling is performed for a reduction of the volume size and two fully connected (FC) layers. Each one follows a softmax classifier with 4096 nodes.

1.3.1 Face mask detection

VGG16 is used to build an efficient network for the detection of face masks. The steps involved in this study are as follows:

• Perform data augmentation to get a clear view of an image sample from a different angle. Keras ImageDataGenerator function is used with rescaling, zoom range, horizontal flip, and shear_range for the proposed work.

• Used a pertained ImageNet to fine-tune the VGG16 architecture.

• Design an FC layer and then finally load the input data.

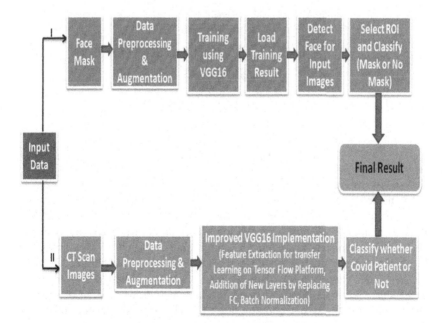

FIGURE 1.2 Proposed work block diagram.

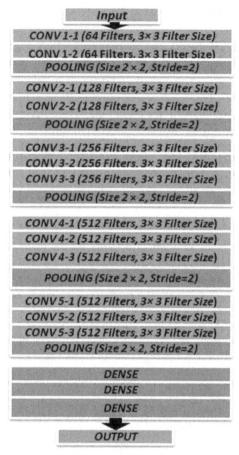

FIGURE 1.3   VGG16 architecture (Simonyan and Zisserman, 2014).

When loading the input image, it resizes to 224 × 224 pixels as a preprocessing. For object detection, face detection using haar-based cascade classifiers is an efficient method proposed by Paul Viola and Michael Jones in 2001 in their paper (Viola and Jones, 2001). In this machine learning technique, a cascade function is learned from both negative and positive images. It is then applied to other images for object detection. Then we need to derive haar characteristics from it. Each feature is a single value produced by subtracting the total of pixels in the white and black rectangles. If faces are identified, the positions of the detected faces are returned as Rect(x, y, w, h). Once we have these positions, we can create a face region of interest (ROI) and apply mask detection on that ROI. The proposed layered architecture of the VGG16 is shown in Fig. 1.4.

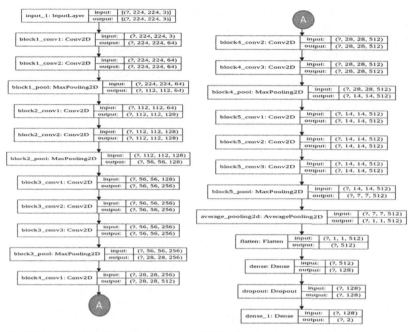

FIGURE 1.4 Proposed layered architecture of VGG16 for face mask detection.

1.3.2 Detection of COVID from CT images

The objective is to train advanced VGG16 deep convolutional model for findings of CT image patterns (vascular dilation, ground glass, traction bronchiectasis, crazy paving, architectural distortion, subpleural bands, etc.) and to classify whether a person belongs to the CT_COVID class or CT_NonCOVID class. Thus by simply uploading images of the CT scan of a patient, the proposed research would assist in the specific preliminary screening of those deemed to be positive for the deadly virus so that clinical testing can follow. Fig. 1.5 shows some sample images from CT images dataset.

The proposed work uses the concept of data augmentation, dropout, normalization, and transfer learning principle for detection. Fig. 1.6 shows the layered architecture of the VGG16 for detection of COVID from CT images. Data augmentation enriches the training data by creating new examples by transforming existing ones by random. This way we increase the size of the training set artificially, reducing overfitting. Techniques for augmenting data such as padding, cropping, and horizontal flipping are widely used to train large neural networks. In the transfer learning method, a developed model is reused for the other task.

Dropout is a regularization method that approximates the training of a large number of neural networks in parallel with various architectures.

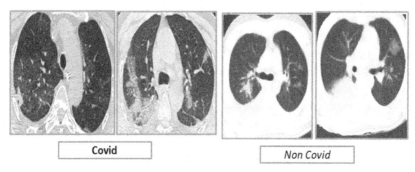

FIGURE 1.5 Sample images from CT images dataset. *CT*, Computed tomography.

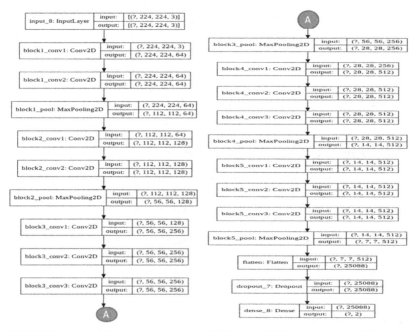

FIGURE 1.6 Proposed layered architecture of VGG16 for COVID detection from CT images. *CT*, Computed tomography.

During training, several layer outputs are randomly "dropped out" or over-looked. Batch normalization is a method used to train a deep CNN, which standardizes the inputs into one layer for each mini-batch. This results in a stabilization of the learning process and a drastic reduction of training epochs number needed for deep networks.

1.4   Results and analysis

The model is trained on Intel Core i5 6th Generation, 2.30 GHz CPU, 12 GB RAM and 2 GB of AMD Radeon R5 M330 graphics engine support on Windows 10 operating system. Accuracy curve, loss curve, confusion matrix, and classification report (precision, recall, f1 score) based analysis is performed for both the experiments. Predictions were generated by running the trained models on images of the test set. The following equation shows the mathematics behind the work.

Accuracy is the fraction of predictions that our model has been accurate.

$$\text{Accuracy} = \frac{\text{Number of correct predictions}}{\text{Total number of predictions made}} \quad (1.1)$$

Accuracy can be measured for binary classification in terms of positives and negatives as follows:

$$\text{Accuracy} = \frac{(TP + TN)}{(TP + TN + FP + FN)} \quad (1.2)$$

where TP = True Positives, FP = False Positives, TN = True Negatives and FN = False Negatives.

When dealing with Log Loss, the classifier will assign each class a probability for all samples. If there are N samples of the groups M, then the Log Loss is determined as follows:

$$\text{Logarithmic Loss} = \frac{-1}{N} \sum_{i=1}^{N} \sum_{j=1}^{M} y_{ij} * \log(p)_{ij} \quad (1.3)$$

Precision: It is the ratio of relevant instances to the retrieved instances.

$$\text{Precision} = TP/(TP + FP) \quad (1.4)$$

Recall: It is the ratio of relevant instances to the total amount of relevant instances that were retrieved.

$$\text{Recall} = TP/(FN + TP) \quad (1.5)$$

F1 Score: It is the weighted average of Precision and Recall.

$$\text{F1 Score} = 2 \times (\text{Precision} \times \text{Recall})/(\text{Precision} + \text{Recall}) \quad (1.6)$$

1.4.1   Face mask detection

This experiment has been performed by looking without a mask or with a mask on real-time faces. In the dataset there are 1376 images, 1104 of which have been used for training and 272 for testing. Fig. 1.7 shows some sample images of masked and without masked faces with detection of face using haar-based cascade classifiers. During training, the total parameters are

FIGURE 1.7 Sample images of masked and without masked faces with detection of face.

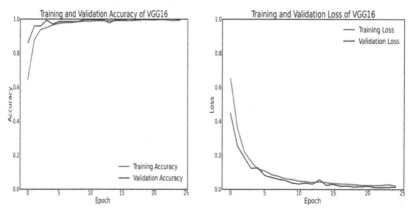

FIGURE 1.8 Accuracy and loss curve for face mask detection.

14,780,610 out of which 65,922 are trainable parameters and 14,714,688 nontrainable parameters.

The proposed work recorded training accuracy of 99.64% with the loss of logarithm 0.02 and validation accuracy of 99.63% with the loss of logarithm 0.02 in only 25 epochs with a batch size of 32 images. For the experiment, Precision, Recall and F1-Score are recorded 99.26%, 100%, and 99.63%, respectively. Fig. 1.8 displays the accuracy curve and loss curve for this experiment. Adam optimizer is used, which measures the individual adaptive learning rate for each parameter from first and second gradient moment's estimates. Confusion matrix-based analysis is done to test the results as shown in Fig. 1.9.

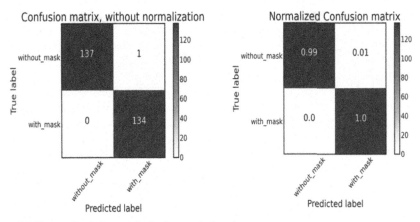

FIGURE 1.9 Confusion matrix for face mask detection.

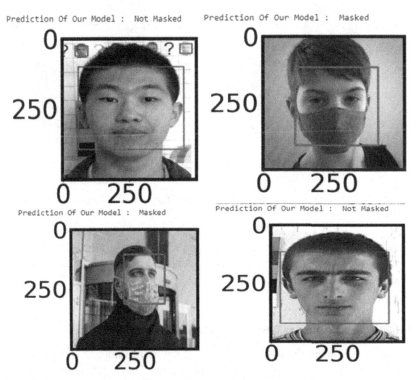

FIGURE 1.10 Prediction of random input image by the proposed model.

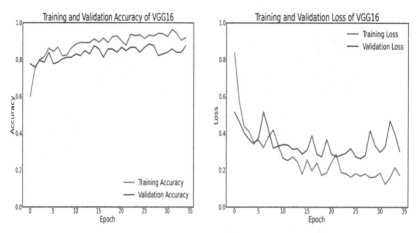

FIGURE 1.11 Accuracy and loss curve for COVID detection from CT scan images. *CT*, Computed tomography.

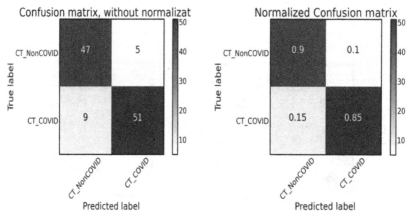

FIGURE 1.12 Confusion matrix for COVID detection from CT scan images. *CT*, Computed tomography.

Fig. 1.10 shows the true and predicted class of any arbitrary size at a given input image. It also reflects faces identified inside a bounding rectangle with their respective pixel-level image.

1.4.2 CT scan image-based COVID-19 patient identification

For this experiment, the dataset contains a total of 746 images, 634 of which have been used for training and 112 for testing. During training, the total parameters are 14,764,866 out of which 50,178 are trainable parameters and

14,714,688 nontrainable parameters. The proposed work recorded training accuracy of 97.63% with the loss of logarithm 0.07 and validation accuracy of 87.50% with the loss of logarithm 0.30 in only 35 epochs with a batch size of 32 images. Fig. 1.11 displays the accuracy curve and loss curve for the proposed model. For the experiment, overall Precision, Recall and F1-Score are recorded 91.07%, 85.00%, and 87.93%, respectively. The confusion matrix for the experiment is shown in Fig. 1.12.

1.5 Conclusion

With COVID-19, the 2020 era began and the image of these viruses is very pathetic. So far no such specific medicine has been invented and that would be the main cause of concern. Two models have been introduced in this proposed work that can be used to identify face masks and verify whether or not a patient with COVID-19 has been compromised by X-ray images, which also contribute to public healthcare. The proposed work recorded a validation accuracy of 99.63% with logarithm loss 0.02 in only 25 epochs for face mask detection and validation accuracy of 87.50% with logarithm loss 0.30 in only 35 epochs for COVID detection from CT images using the VGG16 model. Therefore AI-based face mask detection and radiography applied to CT images are a valuable prevention measure and it proves the technological viability and general applicability of a deep learning approach to enable implementation of preventive measures for global pandemic COVID-19. This technical approach will help researchers find the best possible solution to resolve the future situation.

References

Ai, T., Yang, Z., Hou, H., Zhan, C., Chen, C., Lv, W., & Xia, L. (2020). Correlation of chest CT and RT-PCR testing in coronavirus disease 2019 (COVID-19) in China: A report of 1014 cases. *Radiology*, 200642.

Chung, M., Bernheim, A., Mei, X., Zhang, N., Huang, M., Zeng, X., & Jacobi, A. (2020). CT imaging features of 2019 novel coronavirus (2019-nCoV). *Radiology*, *295*(1), 202−207.

Chu, D. K., Akl, E. A., Duda, S., Solo, K., Yaacoub, S., Schünemann, H. J., & Hajizadeh, A. (2020). Physical distancing, face masks, and eye protection to prevent person-to-person transmission of SARS-CoV-2 and COVID-19: A systematic review and meta-analysis. *The Lancet*.

Ejaz, M.S., Islam, M.R., Sifatullah, M., & Sarker, A. (2019). Implementation of principal component analysis on masked and non-masked face recognition. In *2019 1st International Conference on Advances in Science, Engineering and Robotics Technology (ICASERT)* (pp. 1−5). IEEE.

Fang, Y., Zhang, H., Xie, J., Lin, M., Ying, L., Pang, P., & Ji, W. (2020). The sensitivity of chest CT for COVID-19: Comparison to RT-PCR. *Radiology*, 200432.

Fan, D. P., Zhou, T., Ji, G. P., Zhou, Y., Chen, G., Fu, H., & Shao, L. (2020). Inf-Net: Automatic COVID-19 lung infection segmentation from CT images. *IEEE Transactions on Medical Imaging*.

Guo, Y. R., Cao, Q. D., Hong, Z. S., Tan, Y. Y., Chen, S. D., Jin, H. J., & Yan, Y. (2020). The origin, transmission and clinical therapies on coronavirus disease 2019 (COVID-19) outbreak—An update on the status. *Military Medical Research, 7*(1), 1–10.

Han, Z., Wei, B., Hong, Y., Li, T., Cong, J., Zhu, X., & Zhang, W. (2020). Accurate screening of COVID-19 using attention based deep 3D multiple instance learning. *IEEE Transactions on Medical Imaging.*

Kumar, K., & Shrimankar, D. D. (2017). F-DES: Fast and deep event summarization. *IEEE Transactions on Multimedia, 20*(2), 323–334.

Kumar, K., Shrimankar, D. D., & Singh, N. (2018). Eratosthenes sieve based key-frame extraction technique for event summarization in videos. *Multimedia Tools and Applications, 77*(6), 7383–7404.

Kumar, K. (2019). EVS-DK: Event video skimming using deep keyframe. *Journal of Visual Communication and Image Representation, 58*, 345–352.

Leung, N. H., Chu, D. K., Shiu, E. Y., Chan, K. H., McDevitt, J. J., Hau, B. J., & Seto, W. H. (2020). Respiratory virus shedding in exhaled breath and efficacy of face masks. *Nature Medicine, 26*(5), 676–680.

Li, S., Ning, X., Yu, L., Zhang, L., Dong, X., Shi, Y., & He, W. (2020). Multi-angle head pose classification when wearing the mask for face recognition under the COVID-19 coronavirus epidemic. In *2020 International conference on high performance big data and intelligent systems (HPBD&IS)* (pp. 1–5). IEEE.

Li, C., Wang, R., Li, J., & Fei, L. (2020). *Face detection based on YOLOv3. Recent trends in intelligent computing, communication and devices* (pp. 277–284). Singapore: Springer.

Loey, M., Manogaran, G., Taha, M. H. N., & Khalifa, N. E. M. (2020). A hybrid deep transfer learning model with machine learning methods for face mask detection in the era of the COVID-19 pandemic. *Measurement*, 108288.

MacIntyre, C. R., & Chughtai, A. A. (2020). A rapid systematic review of the efficacy of face masks and respirators against coronaviruses and other respiratory transmissible viruses for the community, healthcare workers and sick patients. *International Journal of Nursing Studies*, 103629.

Sethy, P.K., & Behera, S.K. (2020). Detection of coronavirus disease (COVID-19) based on deep features. Preprints, 2020030300, 2020.

Simonyan, K., & Zisserman, A. (2014). Very deep convolutional networks for large-scale image recognition. arXiv preprint arXiv:1409.1556.

Viola, P., & Jones, M. (2001). Rapid object detection using a boosted cascade of simple features. In *Proceedings of the 2001 IEEE Computer Society Conference on Computer Vision and Pattern Recognition. CVPR 2001* (Vol. 1, pp. I-I). IEEE.

Wang, X., Deng, X., Fu, Q., Zhou, Q., Feng, J., Ma, H., & Zheng, C. (2020). A weakly-supervised framework for COVID-19 classification and lesion localization from chest CT. *IEEE Transactions on Medical Imaging.*

Wang, L., & Wong, A. (2020). COVID-Net: A tailored deep convolutional neural network design for detection of COVID-19 cases from chest X-ray images. arXiv preprint arXiv:2003.09871.

World Health Organization (WHO) in 2020. *WHO coronavirus disease (COVID-19) dashboard.* Retrieved from https://covid19.who.int/. (Accessed 15 2020).

Chapter 2

Internet of Things-based smart helmet to detect possible COVID-19 infections

Chanchal Ahlawat and Rajalakshmi Krishnamurthi
Department of Computer Science and Engineering, Jaypee Institute of Information and Technology, Noida, India

2.1 Introduction

Presently, the world faced COVID-19, a new disease of the coronavirus family. A new virus was announced by the World Health Organization (WHO) named as 2019-nCoV in January 2020 (World Health Organization, 2020a). The world has detected a large number of COVID-19 cases from December 2019 onward. As of now there is no vaccine for COVID-19, but the recovery rate without any treatment is about 80% (Hageman, 2020). People with low immunity, old age, and medical problems, especially related to the lungs, are more prone to COVID-19 disease. COVID-19 symptoms are much similar to the symptoms of normal flu. In normal flu cough, cold, and breathing problems are common. The recovery rate of a person suffering from COVID-19 can take 14−16 days due to its incubation period of 14 days. As all of us know that prevention is better than cure; therefore for prevention of COVID-19 one should wash hands frequently, avoid touching the mouth, nose, and face, and avoid any type of physical contact with others, that is social distancing should be kept in mind (1 meter or 3 feet). COVID-19 has been declared a pandemic by the WHO. Till now there is no vaccine available in the market. But according to different studies, it has been found that this virus is susceptible to ultraviolet rays and heat (Cascella, Rajnik, Cuomo, Dulebohn, & Di Napoli, 2020). Therefore all the preparations regarding medical facilities from hospitals to the masks, PPE kits, and ventilators should be adequate worldwide. As the number of patients is increasing at a great pace, the need for all these medical facilities is in priority to make it possible for health workers to operate efficiently. To deal with COVID-19 one has to know the accurate growth rate of COVID-19 cases of other countries and

Cyber-Physical Systems. DOI: https://doi.org/10.1016/B978-0-12-824557-6.00004-2
15

which country has the highest growth rate, what is the reason behind this, and how they are handling the current situation? It can help the medical and administrative authorities.

The origin of this deadly virus is situated in China's city called Wuhan. Wuhan has a wholesale market especially for sea foods and other animals and this virus came from the intake of one of these animals (Singhal, 2020). Initially it spread to over 37 countries globally, with around 8000 confirmed cases and about 774 death cases (Centers for Disease Control Prevention, 2017). According to the WHO, the total number of confirmed cases world-wide is about 13,876,441cases, and 593,087 are the total number of deaths till July 17, 2020. COVID-19 wreaked havoc in many regions, the United States stands at 7,306,376 confirmed cases, Europe stands at 3,042,330 confirmed cases, and the East Mediterranean region stands at1,360,791 confirmed cases (https://COVID19.who.int/?gclid = Cj0KCQjwu8r4BRCzARIsAA21i_DGpHh9yJeQSHGLxoPJI0rhmCYgtrYKdA4BcXJQpmFVtAUUJQcGbUaAoWhEALw_wcB).

To overcome the growing rate of COVID-19 pandemic several actions have been taken worldwide. All the countries are doing their best to deal with this virus. For example India's state and central governments have taken various strict actions earlier like imposing lockdowns. One of the vital actions taken by the government of Delhi is by having 5T plan as shown in Fig. 2.1.

In the current pandemic circumstances, countries worldwide are trying to maneuver to fight against COVID-19 and to obtain a fixed and cost-effective solution to get rid of this. The situations are getting worse day by day even after taking various vigorous steps to control this deadly virus. This situation is the same worldwide. In Fig. 2.2 the condition of active cases from the months of February to June 2020 are shown. Therefore the requirement of technology arises to get control over this virus by descrying the infectious body. One such technology is "Internet of Things" (IoT), which can interconnect various physical devices such as sensor, actuators, mobile, etc and allow them to interact with each other over a defined network and requires no human intervention during the whole process. All devices are allied with a unique identification code (Bai, Yang, Wang, Tong, & Zhu, 2020; Haleem, Javaid, & Khan, 2019). IoT has already accustomed to the various domains for accomplishing different motives like smart home, smart power grid, smart vehicles, industries, and healthcare department. As IoT can facilitate in the healthcare sector, it can be used to get rid of or to handle COVID-19 (Singh, Javaid, Haleem, & Suman, 2020). The key contributions of this chapter are as follows:

- To discuss the necessity of the IoT-based solutions approach toward handling COVID-19.
- This chapter describes the in-depth study on technology of IoT-based devices and IoT-based smart helmet to detect the infectee of COVID-19.

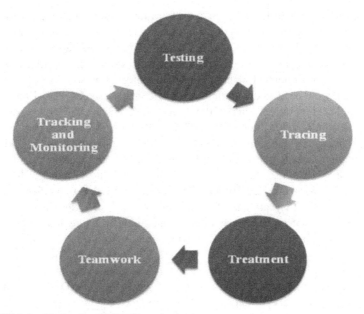

FIGURE 2.1   5T plan for COVID-19 pandemic.

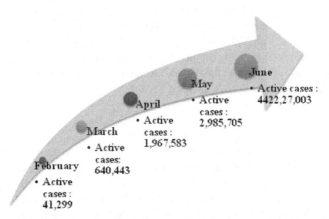

FIGURE 2.2   Timeline of total number of active infected cases from February to June 2020 worldwide.

- Also, it discusses about the future possibilities and applications of IoT-based devices in the healthcare sector.

2.1.1   Epidemiology

At the end of the year December 2019, a virus was reported at the "Wuhan city" of China, amidst people associated with the local seafood market

(Adhikari et al., 2020; Li et al., 2020). After that various researches started to know the severity, characteristics, effectiveness, and transmission capabilities of this virus (CDC, 2020). It came out that the initial cases had its connection with the wet market (seafood market) and later it was found that transmission of the virus was through close contact with the infected person. Various people of the healthcare department got infected while treating the COVID-19 patients (Li, Wei, Li, Hongwei, & Shi, 2020; Liu et al., 2020; Medical Expert Group of Tongji Hospital, 2020; Zhou et al., 2020). Also it was found that people with immunodeficiency and normal people were both prone to this virus, and age distribution between young and old was 25−89 years. Children and even infants cases were also detected (Medical Expert Group of Tongji Hospital, 2020; Wang & Wang, 2020). Further studies showed the median age of patients between 15 years and 89 years was 59 years. The incubation period means the time interval between when an individual got infected for the first time and when he/she shows the first symptoms. According to Chinese health authorities, the patient has 7 days of an average incubation period from 2 to 14 days (National Health Commission of People's Republic of China, 2020). On January 2020, China had 11,791 confirmed cases. Soon COVID-19 blast affect was shown in every country and part of the world (National Health Commission of People's Republic of China, 2020). According to the WHO Emergency Dashboard, the situation of COVID-19 till July 22, 2020 was that there have been 14,562,550 confirmed cases of COVID-19, including 607,781 deaths. Outside China the situation of confirmed cases in different countries are as follows: the United States (374,248), Brazil (2,098,389), India (1,155,191), Russia Federation (783,328), South Africa (373,628), Peru (353,590), Mexico (344,224), Chile (333,029), the United Kingdom (295,3760), Iran (276,202), Pakistan (266,096), and Spain (264,836).

2.1.2 Treatment

Due to the outbreak of the coronavirus, the whole world tends to be in an unprecedented state of austere confusion. Till now, there is neither specific treatment for the virus found nor any preventive vaccine invented to prevent the coronavirus. COVID-19 can be symptomatic or asymptomatic; therefore treatment depends on the symptoms shown by the individual patient (Chamola, Hassija, Gupta, & Guizani, 2020). Some patients having asymptomatic behavior require no treatment and can recover on their own. While patients with minor symptoms can go for home isolation and require minor medications. However, patients with critical conditions such as breathing disorder, asthma, low immune system, etc. need to be hospitalized. In conditions like hypoxemia patients require extra oxygen supply or ventilators. Various antibiotics and antifungals are provided to the patients according to the situations. Therapy called renal replacement can be required if kidney patients suffer from the coronavirus (Wang, Hu, et al., 2020).

Due to the severity of COVID-19, various researches are being made to control the virus as soon as possible. Some of the potential vaccines and drugs are—Moderna's Mrna-1273 (Moderna, 2020; Park, 2020), Pittcovacc (Pittsburgh Coronavirus Vaccine) (UPMC, 2020; Kim et al.), Johnson & Johnson's COVID-19 lead vaccine (Johnson & Johnson, 2020). Drugs such as hydroxychloroquine and arbidol have shown good effects on the patients, and many hospitals are undergoing the trial of these drugs (Holshue et al., 2020; Wang, Cao, et al., 2020). Similarly, favipiravir, bromhexine with hydroxychloroquine, remdesivir, ritonavir, etc. are drugs which are in the different phases of the clinical trial [Source: United States National Library of Medicine (trial phase as on April 29, 2020)]. There are around 99% COVID-19 patients from Wuhan. 76% of patients require oxygen, 13% need noninvasive ventilation, 4% need mechanical ventilation, 9% require continuous renal-replacement therapy, 71% require antibiotics, and 15% require antifungal drugs (Chen, Zhou, & Dong, 2020; Singhal, 2020).

Technologies such as the IoT has been used by researchers to treat COVID-19. Bai et al. (2020) introduced a system called "COVID-19 Intelligent Diagnosis and Treatment Assistant Program" (nCapp) for diagnosis and treatment of COVID-19 patients. The nCapp has three levels "Comprehensive Perception," "Reliable Transmission," and "Intelligent Processing" of the IoT. Fifteen questionnaires have been made for the diagnosis and detection purpose, and based on that patients marked as confirmed or suspected are automatically sent to the nearby doctors. The doctors use a smartphone with nCapp assistant software to connect with the three levels of the cloud. Doctors or physicians have eight functions to manage different activities, that is register, consultations, treatment, map, diagnosis, specialist, protection, and information. nCapp also suggests some treatment to the patient. nCapp system is capable to treat the patient in their home itself and can avoid transmission and prevention (Bai et al., 2020).

2.1.3 Prevention

Prevention of any disease is an essential measure to take. Since till now no specific treatment is available to control this infectious disease, the prevention of COVID-19 is a difficult task. It may be symptomatic or asymptomatic and can spread from the day the individual got infected or even after recovery from the virus (Singhal, 2020). Medical institutes like WHO, Centers for Disease Control and Prevention (CDC) issued some guidelines to prevent this virus (Centers for Disease Control and Prevention, 2020; World Health Organization, 2020b). They suggest to avoid visiting the containment zones (higher risk of COVID-19), try to maintain distance from symptomatic individuals, and also to avoid the nonvegetarian items from the higher risk zones.

Various research has been undergoing to control this pandemic but some researchers and authors also provide prevention techniques (for different fields) or applications. A chat boot (Beboot) has been introduced by a Japanese company based on artificial intelligence that will hand over all the updated information about the outbreak of coronavirus (Bespoke, 2020). Basile et al. (2020) also defined the preventive measures for the hemodialysis centers. As hemodialysis centers' patients are more prone to infection so extra care or preventive measures have to be taken. According to a report (Ma, Diao, & Lv, 2019), Wuhan city of China had 37 cases out of 233 in hemodialysis centers from 14 January to 17 February, among which four staff members out of 33 were found infected. The most effective preventive measure is the isolation of hemodialysis patients from COVID-19 patients.

Therefore isolation is the most appropriate preventive measure of COVID-19 and control of infection that includes effective measures at the time of diagnosis and provisioning of infected patient (CDC, 2020). Fig. 2.3 shows some preventive measures for COVID-19.

To prevent this pandemic the first thing that we have to do is to limit the transmission of cases. To limit the cases some preventive measures are introduced by various standard organizations such as the WHO, which are as follows (Cascella et al., 2020):

- Avoid the direct contact from the person who has symptoms like fever, cold, breathing problem.
- Follow social distancing.
- Use of PPE kit.
- Wash your hands frequently, or sanitize whenever going outside.
- Wear mask whenever going outside and proper disposal of used mask.
- Avoid unprotected contact with farm or wild animals.
- Try to improve your immune system.
- Cover the face with tissue/handkerchief while sneezing or coughing and dispose off properly.
- Can cover your face with flexed elbow while sneezing (CDC, 2020).

These preventive measures should be followed by everyone, should carry hand sanitizer having 80% alcohol, avoid touching your face, eyes, nose

FIGURE 2.3 Preventive measures for COVID-19.

frequently if you are going outside into the environment or in containment zones. Proper use of PPE kit containing N95 mask, gloves, face shield, and gown by the healthcare department (Cascella et al., 2020).

Various technologies also help us to prevent the COVID-19 in enormous ways. One of the technologies like IoT has been adopted in various fields to make things smart like smart home automation, smart grid, and smart healthcare system. Although IoT has its effects on the healthcare system, it is necessary to have IoT to deal with this pandemic. Therefore an idea for IoT-based smart helmet proves to be a promising IoT-enabled device to prevent the coronavirus transmission and detection of infectee becomes easier.

2.1.4   Symptoms

Symptoms of COVID-19 may vary from individual to individual. Some may show asymptomatic behavior (Singhal, 2020). Some situations may arise where an individual has no symptoms but suffering from acute respiratory distress syndrome and multiorder dysfunction syndrome (Chamola et al., 2020). The WHO along with China collaboration and about 55,924 labs found similar clinical characteristics in COVID-19 patients such as fever, cough, and fatigue whereas some cases have symptoms like headache, sore throat, and breathing difficulty. Some symptoms are found rarely in a patient like diarrhea, nausea, and nasal congestion. People having a medical history of diabetes, asthma, hypertension, etc. are more prone to COVID-19 (Centers for Disease Control Prevention (CDC), 2020). Table 2.1 gives a view of the types of symptoms with the percentage of having these symptoms in the infectee.

2.1.5   Stages of COVID-19

According to the WHO, the pandemic COVID-19 has four stages of transmission in every country worldwide and does not show any variations in the stages of transmission (Chamola et al., 2020; WHO, 2020c). With the help of these stages, other countries try to implement preventive measures to deal with this pandemic, the most appropriate example of this being a lockdown. Fig. 2.4 consists of four transmission stages of COVID-19. The four stages description is as follows:

- Imported cases only—This is the first stage of COVID-19. In this stage, only few people got infected by this virus, based on their travel history to the infected area (WHO, 2020d).
- Local transmission—The second stage of COVID-19 exists when there is erratic infection of the virus between the people. Contact tracing becomes easier as the infection came from nearby people such as neighbors,

TABLE 2.1 Symptoms of COVID-19.

S. no	Types of symptoms	Symptoms	Percentage (%)
1	Most common symptoms	• Fever	87.9
		• Dry cough	67.7
		• Fatigue	38.1
		• Sputum production	33.4
2	Less common symptoms	• Breathing disorder	18.6
		• Sore throat	13.9
		• Headache	13.6
		• Chills	11.4
		• Myalgia	14.8
3	Rare symptoms	• Nausea	5.0
		• Diarrhea	3.7
		• Hypostasis	0.8
		• Conjunctional congestion	0.9

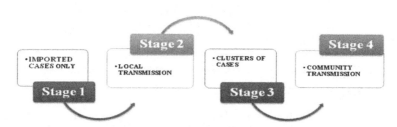

FIGURE 2.4 Four transmission stages of COVID-19.

family, relatives, colleagues, etc. A proper quarantine process can be done easily.

• Clusters of cases—The third stage in any country is considered when some group cases persist in a particular zone in a different area of the country. Tracing the contact of the individual patient becomes hard as the infectee does not have any history of travel or not even have contact with the other infected. The area where such a group of cases are found has to be converted into a containment zone.

- Community spread—The last stage of COVID-19 is the stage of community spread, that is extreme increase in the COVID-19 cases with an increase in the number of deaths. The outbreak of the patients becomes out of control and no prevention techniques seem to be effective. Vaccination is the only way to get control over this outbreak. Community spread like situation can be seen in various countries such as the United States, Turkey, and Canada (WHO, 2020c).

2.1.6   Key merits of IoT for COVID-19 pandemic

IoT is one of the technologies that can be used in every field. When it comes to the healthcare sector IoT is already being used by various hospitals. Now when the world is facing a COVID-19 outbreak there is a need to have an innovative technology that helps us to deal with COVID-19. IoT can assure that infectees are in quarantine. With the help of a proper IoT-based monitoring system we can monitor the infectees. IoT can enhance the efficiency of all the workers with essential duties such as doctors, nurses, police, and other medical staff by declining their workload. Various applications or devices have been developed to control or to prevent the spread of COVID-19, which is discussed in a later section.

Using IoT for COVID-19 can deduct the possibility of mistakes, lower the expenses, provide superior treatment, meliorate diagnosis, effectual control, and able to measure the blood pressure, heart rate, and sugar levels (Mohammed, Syamsudin, et al., 2020; Singh, Javaid, et al., 2020; Vaishya, Javaid, Khan, & Haleem, 2020).

2.1.7   Internet of Things process required for COVID-19

Due to the various challenges in the way to control, prevent, or diagnose COVID-19, IoT seems to be a very innovative technology. IoT is capable to sense the real-time data efficiently, which can help to get or to sense the real-time data of the COVID-19 infected patient and able to gather all the required information. Also, it collects the real-time information of various areas during the lockdown period (Allam & Jones, 2020; Dewey, Hingle, Goelz, & Linzer, 2020; Javaid, Vaishya, Bahl, Suman, & Vaish, 2020; Singh, Javaid, et al., 2020).

The process to grapple with COVID-19 using IoT consists of four steps. In the initial step, IoT is used to collect the health data or to collect the information regarding the crowd during lockdown via drones. In the second step, collected information can be managed by using virtual management. In the third step, analysis of the received data is done to take necessary actions. In the final step, a report will be generated, and timely follow-up has to be done (Gupta, Abdelsalam, & Mittal, 2020; Stoessl, Bhatia, & Merello, 2020).

2.1.8 IoT applications for COVID-19

The areas of application of IoT are very vast. There are different applications of IoT used in the healthcare sector. IoT consists of a huge number of interconnected devices and forms a network that can be used in the healthcare sector for the health management system. It can trace the disease of the patient by using any wearable like fitness band, which is able to get the temperature, heartbeat, etc., and can send alerts or messages to their concerned doctors or hospital. All this information can be collected without any human intervention and can use this data for analysis or for making the decision for the treatment of the patient (Ghosh, Gupta, & Misra, 2020; Gupta & Misra, 2020; Gupta, Ghosh, Singh, & Misra, 2020; Singh, Javaid, et al., 2020; Yang, Gentile, Shen, & Cheng, 2020; Zheng et al., 2020). IoT is capable to predict the imminent condition by using the sensed data. Various fraud claims can be made by the fraudsters to verify the details; hence insurance companies can also make use of this technology. Table 2.2 provides the details of the IoT applications with its usage.

2.2 Related work

Although the survey on COVID-19 is not very vast as it was found in December 2019, medical teams of various countries are trying to find out the solution to this virus but have not succeeded yet. But researchers are also giving their contribution.

Dong and Du (2020) present an online interactive dashboard to envision and track the overall cases of COVID-19 in real-time. On January 22, the dashboard has been shared to show the information about the current situations, that is number of deaths, the number of infected patients, and number of recoveries. And it is freely available online. From January 22−31, all the collection of data and updating is done manually twice a day, that is morning and night. A platform called DXY is used by the Chinese government and medical staff.

Maanak Gupta et al. (Khanna and Anand, 2016) presented an architecture and some use case of COVID-19 to implement social distancing using smart city and intelligent transport system. The architecture contains physical devices like smart transport, sensor devices, and smart traffic light system, responsible for exchanging data with nearby smart devices and at the same time sends data to the cloud for further processing for communication Message Queuing Telemetry Transport protocols. To implement social distancing a drone can be used to track the different areas and if in any community social distancing is not followed then messages are sent to the nearby traffic controllers. So that action can be taken. Similarly, some use cases are defined by the author to maintain social distancing like monitoring large gatherings, rerouting traffic to reduce footprint, smart parking (Khanna &

TABLE 2.2 The IoT applications for healthcare sector (Bai et al., 2020; Hassen, Ayari, & Hamdi, 2020; Singh, Javaid, et al., 2020; Swayamsiddha & Mohanty, 2020).

S. no.	IoT applications	Usage
1	Inter connecting hospital via IoT.	To fulfill the requirement of integrated network of hospital using IoT.
2	Transparent COVID-19 treatment.	To get the treatment of COVID-19 pandemic in a fair manner.
3	Smart tracing of infected patients,	To trace the COVID-19 patients in the starting days of infection.
4	Connect all medical tools and devices through the internet.	To transmit the real-time treatment information, IoT connection of medical tools and devices via internet is necessary.
5	Accurate forecasting of virus.	To predict the current situations some statistical methods are used so that data can be helpful for medical teams for the treatment of COVID-19 patient.
6	Wireless healthcare network to identify COVID-19 patient.	To identify the patient infected with COVID-19 by using various authentic applications in smart phones.
7	Rapid COVID-19 screening.	To diagnose the infection at the very initial stage by using the smart devices in IoT environment.
8	IoT-based home hospitalization system.	To treat the disease at home by monitoring patient information via mobile applications by the doctors.
9	Application of cognitive Internet of Medical Things.	For tracing patient in real-time, prevention, remote monitoring, rapid diagnosis, screening, and surveillance.
10	COVID-19 Intelligent Diagnosis and Treatment Assistant Program (nCapp).	Diagnosis of the patient in real-time and make treatment available by sending the data to the doctors.

Anand, 2016), intelligent transport system, and big data and artificial intelligence.

Singh, Javaid, et al. (2020) presented a review of different IoT applications to show the effectiveness of using IoT applications for the COVID-19 pandemic. The author collected the data from different sources like Google Scholar by using different keywords like "IoT," "COVID-19" etc., and also used various blogs. This review is beneficial to handle or to understand the advantages of having IoT applications.

Greco, Percannella, Ritrovato, Tortorella, and Vento (2020) overviewed the trends in the healthcare sector by using the IoT. They discussed how the integration of IoT and cloud computing leads to sustaining real-time applications. By sensing the real-time data from a wearable sensor, the processing and analysis of the received enormous data help to obtain the real condition of the patient. The authors reviewed various papers on IoT solution in the field of healthcare, initially from sensing the patient via wearable sensors to the latest trends for smart health by suing emerging IoT technology called Fog and edge computing.

Kummitha (2020) presented a literature survey to show the discrepancy between practices to control the COVID-19 pandemic. The author analogizes two contrasting approaches, that is human-driven approach and techno-driven approach to know which approach shows effective preventive measures for COVID-19. The techno-driven approach was adopted by China while the human-driven approach was followed by the western democracies. Literature focuses on understanding the relationship between humans and technology, which provides some observations for prevention and control of COVID-19.

Wang, Sun, Duong, Nguyen, and Hanzo (2020) presented a scheme to diagnose infected cases of COVID-19. The authors used a weighted undirected graph to propose the social IoT network topology evolving during the outbreak of a pandemic. They proposed a framework for the minimum-weight vertex cover problem of graph theory (He et al., 2017) and an algorithm for risk-aware adaptive identification. Simulation has been performed on a realistic dataset to show the suppression of transmission in a large and small area.

Angurala, Bala, Bamber, Kaur, and Singh (2020) proposed a system called Drone-based COVID-19 Medical Service (DBCMS), especially for the healthcare workers as they are susceptible to COVID-19 infection while treatment of the infected individuals. To control or prevent the infection, the drone plays an important role in the proposed mechanism. The author showed the overall infected healthcare workers till April 22, 2020 (https://www.thehindu.com/data/how-many-doctors-and-nurses-have-tested-positive-for coronavirus-in-india/article31410464.ece). DBCMS architecture consists of three layers. The first layer gathers the samples, second stage is responsible for the serious patients that require immediate consultation and medication by the expert doctors. Stage three is for the top-level authority to alert them for the alarming COVID-19 situations.

Chamola et al. (2020) provided a detailed review of COVID-19. The main aim is to eliminate the false report and to showcase the real facts about the COVID-19 pandemic. They also discussed the technologies such as IoT, artificial intelligence, 5 G, and Machine Learning, which can be useful to control or prevent the COVID-19 outbreak. Various applications also related to these technologies are also mentioned by the authors.

Song, Jiang, Wang, Yang, and Bai (2020) described the role of the IoT for control and prevention of Severe Acute Respiratory Infection by using IoT with AI, sensors, and other network devices. IoT is able to communicate between different hospitals, patients, and other healthcare devices, by which existing medical situations would improve. The authors also discussed the variety of applications in the field of medicine, especially respiratory diseases or infection. They also elaborated on the prevention, development, and prospects of the medical IoT.

2.3 IoT-based smart helmet to detect the infection of COVID-19

2.3.1 Objective

Due to the outbreak of COVID-19, the exigency of the hour is to control and to avert the transmission of the virus. COVID-19 spreads rapidly from individual to individual. All preventive measures are lacking as the thermal screening process to check the temperature of the body takes much time. Also, the people doing the thermal screening process have a higher chance to get infected. Therefore the requisite to do more screening processes at the early stage of the infection and screening process should be done efficiently and by maintaining social distance. Therefore to detect the virus automatically a smart device is introduced, that is "smart helmet."

2.3.2 Methodology

The solution of the IoT is to develop smart helmet and to detect possible COVID-19 infections. IoT applications in the healthcare sector light up the hope for better services and resource availability at the time of requirement for the patients. IoT consists of sensors to sense the biometrics of the patients, big data, and cloud computing for analysis of the data received through telemedicine, and hospital management system. Till now, various researchers are doing their best to use the innovative technologies to cure or prevent the infection, such as monitoring of hand wash via sensor enable devices, health monitoring devices, remote tracking and monitoring of the infectee, etc. (Fong, Wui Yung Chin, Abbas, Jamal, & Ahmed, 2019; Hu, Xie, & Shen, 2013; Mohammed, Desyansah, Al-Zubaidi, & Yusuf; Zamani, Mohammed, & Al-Zubaidi, 2020). Hassen et al. (2020) introduced an application called the hospital management system using IoT to treat the disease at home by monitoring patient information via mobile applications by the doctors. Similarly, Bai et al. (2020) proposed a COVID-19 Intelligent Diagnosis and Treatment Assistant Program (nCapp) that diagnoses the patient in real-time and makes treatment available by sending the data to the doctors. Mohammed, Hazairin, et al. (2020) proposed a system capable to detect and diagnose the COVID-19 using IoT technology. The authors introduced a smart glass to detect the infectious body in the suspected area and send the gathered data to the health officer.

To suspect the infected cases smart helmet is introduced. Smart helmet is fitted with two cameras: thermal camera and normal camera (optical). Thermal cameras do the screening process and are installed on the side of the helmet, which is capable to screen 13 people in 1 minute. An optical camera takes photographs. A sensor device is also installed to the helmet, which is connected to the smart watch to get the temperature of the suspect. Based on the displayed temperature in the smart watch it can be decided if the suspect is COVID-19 positive or not (http://timesofindia.indiatimes.com/ articleshow/77168216.cms?utm_source = contentofinterest&utm_medium = text&utm_campaign = cppst).

2.3.2.1 Efficiency of smart helmet

- Capable of doing mass screening.
- Screening can be done by following social distancing.
- Using IoT can enable to send information to the health officer.
- Capable of storing the infectee data using IoT.

A smart helmet (Mohammed, Hazairin, et al., 2020) has been proposed using IoT solutions, which is capable to automatically detect if the patient is infected by COVID-19 or not using thermal camera. With help of IoT, real-time data are collected after proper screening process.

2.3.2.2 Components of smart helmet

2.3.2.2.1 Thermal camera

A thermal camera also known as an infrared camera uses infrared waves to capture an image similar to the normal camera that exploits light to create an image. It aims to detect the high-temperature bodies by comparing the temperature with other bodies in the containment zone. Therefore when envisaging the body with high temperature it generates the infrared spectra of high intensity.

2.3.2.2.2 Optical camera

An optical camera is a normal camera used to capture the image of the infected or containment zone to use it later for the verification purpose by the concerned authorities.

2.3.2.2.3 Arduino Integrated Development Environment (IDE)

It is a platform written in JAVA. It consists of many characteristics like brace matching, libraries, multiple file compilation, syntax highlighting, etc.

Arduino IDE is loaded up with the Arduino board and uses a one-click process to compile and upload the program. Languages like C and C++ are also supported. It also consists of various input − output methods for software libraries.

2.3.2.2.4   Proteus software

To provide real-time simulation a software is used which is called the Proteus software that offers the simulation, schematics, and circuit design to permit the human to acquire access for the duration of running phase (Jamal, AL Narayanasamy, MohdZaki, & Abbas Helmi, 2019; Mohammed, Syamsudin, et al., 2020).

2.3.2.2.5   Google Location History

Google Location History is the service provided by Google that stores every location of the user visited with each mobile device. The proper management of the data is handled by the Google account and user's routine and mobility can be accessed by the Google Location History (Fong, Chin Wui Yung, Ahmed, & Jamal, 2019).

Smart helmet processing has three modules; all the modules are interconnected with each other. Image processing is the module that processes all the data received from the thermal and optical cameras. Fig. 2.5 shows the design of a smart helmet. The main task of data collection is handled by the smart helmet. The interfacing between the modules happens via the IoT communication links and GSM. The working process of the smart helmet is shown in Fig. 2.6.

Smart helmet is outfitted with two cameras to get the detailed data of body temperature and face detection. This information is provided by the thermal camera and optical camera, respectively. Both the cameras are incorporated in the smart helmet. The process started with the screening process by scanning the area or containment zone by using a thermal camera. After screening of the suspected area and the people, if the temperature is found to

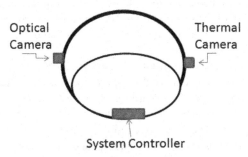

FIGURE 2.5   Design of overall system with system controller, thermal camera, and optical camera.

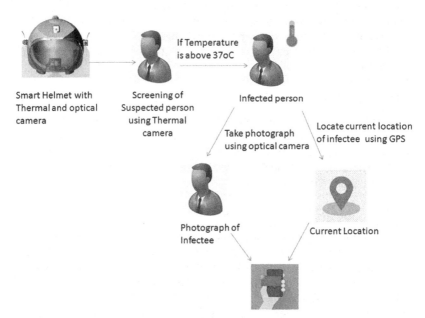

FIGURE 2.6 Working process of the smart helmet.

be higher than 37°C then the chances of COVID-19 infection increases, and the picture will be taken with the help of an optical camera. Now the current location of the infectee or the containment zone is determined by using GPS. Gathered data, that is photographs and location, are sent to the concerned authority and a notification is sent to the mobile phone via GSM with collected information to verify the infectee. A smart helmet mechanism is categorized into three parts. The first part is responsible for the input taken by the two cameras, that is thermal camera and optical camera, and with the application installed on the mobile phone. In the second part processor development is involved in which integration is involved using Arduino DE software. Arduino IDE carries out the coding of the source code and compiles the commands and source code into the NODEMCU V2 processor. The third part is responsible for the output mechanism (Mohammed, Hazairin, et al., 2020).

Cascade classification algorithm is used to face the detection process (Viola & Jones, 2001). Positive and negative images are trained with the use of cascade function in machine learning algorithms. Open CV library used for APIs and cascade object detection that perceive the face of the image is taken by the camera. The face recognition process is a process largely focused for many years. Also, performance and processing from computer vision prospects in video or online streaming become possible in the transferable device. As COVID-19 is a transferable infection, it is necessary to know the history of the infection. Therefore when a suspected person is found to be infected, the history of the visited place needs to be found out. For this, Google Location History was used to obtain a detailed history of the infectee (Ruktanonchai, Ruktanonchai, Floyd, & Tatem, 2018; Sardianos, Varlamis, & Bouras, 2018). Sequential communication used by the Arduino to deliver the collected data, that is infectee body temperature, face recognition, GPS location, NodeMcu a micro-controller have the data and transmit it to the network, to make information available for the worldwide access. Blynk was used as an external server (Bhatnagar et al., 2020; Kumari et al., 2020; Singh, Poonia, et al., 2020).

2.4   Conclusion

The IoT provides a wide-ranging network to the healthcare sector. In the COVID-19 outbreak an infectious and transferable virus hit the world in the last month of 2019 and millions of people have lost their lives. IoT technology provides a way to fasten the preventive and screening process by connecting all devices related to healthcare to the internet and automatically transfer the message to the medical team. IoT make it possible to treat the infected people remotely and can make aware of the impending situations. An innovative smart helmet, integrated with a thermal camera, has come into existence to automatically detect the infectee. Therefore thermography is used, which is a process of thermal screening of an individual or cluster of people. Artificial intelligence and IoT are used to analyze the COVID-19. The diagnosis process using a smart helmet takes less amount of time compared to the normal screening process. In future, applications of IoT-enabled devices can be used to handle this type of pandemic or any other medical emergencies and will able to handle the healthcare sectors or hospitals more efficiently in a cost-effective manner. Therefore it can be concluded that to satisfy the requirement of the healthcare sector to identify, monitor, and to control the COVID-19 remote sensing procedures provides assurance and potential.

References

Adhikari S. P., Meng S., Wu Y. -J., Mao Y. -P., Ye R. -X., Wang Q. -Z., ... Zhoul H. (2020), Epidemiology, causes, clinical manifestation and diagnosis, prevention and control of coronavirus disease (COVID-19) during the early outbreak period: A scoping review. Available from: https://doi.org/10.1186/s40249-020-00646-x.

Allam, Z., & Jones, D. S. (2020). On the coronavirus (COVID-19) outbreak and the smart city-network: Universal data sharing standards coupled with artificial intelligence(AI) to benefit urban health monitoring and management. *Healthcare*, 8(1), 46, Multidisciplinary Digital Publishing Institute.

Angurala, M., Bala, M., Bamber, S. S., Kaur, R., & Singh, P. (2020). An internet of things assisted drone based approach to reduce rapid spread of COVID-19. *Journal of Safety Science and Resilience*, *1*, 31−35.

Bai, L., Yang, D., Wang, X., Tong, L., Zhu, X., et al. (2020). Chinese experts' consensus on the Internet of Things-aided diagnosis and treatment of coronavirus disease 2019 (COVID-19). *Clinical e Health, 3*, 7−15.

Basile, C., Combe, C., Pizzarelli, F., Covic, A., Davenport, A., Kanbay, M., ... Mitra, S. (2020). Recommendations for the prevention, mitigation and containment of the emerging SARS-CoV-2 (COVID-19) pandemic in haemodialysiscentres. *Nephrology, Dialysis, Transplantation: Official Publication of the European Dialysis and Transplant Association - European Renal Association, 35*, 737−741.

Bespoke. (2020). Bebot launches free coronavirus information bot. Available from: https://www.be-spoke.io/index.html.

Bhatnagar, V., Poonia, R. C., Nagar, P., Kumar, S., Singh, V., Raja, L., & Dass, P. (2020). Descriptive analysis of COVID-19 patients in the context of India. *Journal of Interdisciplinary Mathematics*, 1−16. Available from https://doi.org/10.1080/09720502.2020.1761635.

Cascella, M., Rajnik, M., Cuomo, A., Dulebohn, S. C., & Di Napoli, R. (2020). *Features, evaluation and treatment coronavirus (COVID-19) [updated April 6, 2020]'. StatPearls [Internet].* Treasure Island, FL, USA: StatPearls Publishing. 2020. Available from: https://www.ncbi.nlm.nih.gov/books/NBK554776/.

CDC (2020). 2019 novel coronavirus, Wuhan, China. https://www.cdc.gov/coronavirus/2019-nCoV/summary.html. Accessed July 22, 2020.

Centers for Disease Control and Prevention (2020). 2019 novel coronavirus. Available from: https://www.cdc.gov/coronavirus/2019-ncov/about/transmission.html.

Chamola, V., Hassija, V., Gupta, V., & Guizani, M. (2020). A comprehensive review of the COVID-19 pandemic and the role of IoT, drones, AIi, blockchain, and 5G in managing its impact. Special section on deep learning algorithms for internet of medical things, May 26, 2020.

Chen, N., Zhou, M., Dong, X., et al. (2020). Epidemiological and clinical characteristics of 99 cases of 2019 novel coronavirus pneumonia in Wuhan, China: A descriptive study. *Lancet, 395*, 507−513.

UPMC. (April 2020). COVID-19 vaccine candidate shows promise. Available from: https://www.upmc.com/media/news/040220-falogambotto-sars-cov2-vaccine.

Dewey, C., Hingle, S., Goelz, E., & Linzer, M. Supporting clinicians during the COVID-19 pandemic. Annals of Internal Medicine 2020. In press.

Dong, E., & Du, H. (2020), An interactive web-baseddashboard to trackCOVID-19 in real time. Available from: https://doi.org/10.1016/S1473-3099(20)30120-1.

Fong, S. L., Yung, D. C. W., Ahmed, F. Y. H., & Jamal, A. (2019). Smart city bus application with Quick Response (QR) code payment, in ICSCA '19 Proceedings of the 2019 8th *International Conference on Software and Computer Applications*, pp. 248−252, Penang, Malaysia—February 19−21.

Fong, S. L., Chin, D. W. Y., Abbas, R. A., Jamal, A., & Ahmed, F. Y. (2019). Smart city bus application with QR code: A review, 2019 IEEE IEEE *International Conference on Automatic Control and Intelligent Systems* I2CACIS 2019 - Proc., no. June, pp. 34−39.

Ghosh, A., Gupta, R., & Misra, A. (2020). Telemedicine for diabetes care in India during COVID19 pandemic and national lockdown period: guidelines for physicians. *Diabetes & Metabolic Syndrome: Clinical Research & Reviews, 14*(4), 273−276.

Greco, L., Percannella, G., Ritrovato, P., Tortorella, F., & Vento, M. (2020). Trends in IoT based solutions for health care: Moving AI to the edge. *Pattern Recognition Letters, 135*, 346−353.

Gupta, M., Abdelsalam, M., & Mittal, S. (2020). Enabling and enforcing social distancingmeasures using smart city and its infrastructures: A COVID-19 use case. arXivpreprint arXiv:2004.09246. April 13.

Gupta, R., Ghosh, A., Singh, A. K., & Misra, A. (2020). Clinical considerations for patients with diabetes in times of COVID-19 epidemic. *Diabetes & Metabolic Syndrome: Clinical Research & Reviews*, *14*(3), 211e2.

Gupta, R., & Misra, A. (2020). Contentious issues and evolving concepts in the clinicalpresentation and management of patients with COVID-19 infection withreference to use of therapeutic and other drugs used in co-morbid diseases (hypertension, diabetes etc.). *Diabetes & Metabolic Syndrome: Clinical Research & Reviews*, *14*(3), 251e4.

Hageman, J. R. (2020). The coronavirus disease 2019 (COVID-19). *Pediatric Annals*, *49*(3), e99−e100.

Haleem, A., Javaid, M., & Khan, I. H. (2019). Internet of things (IoT) applications in orthopaedics. *Journal of Clinical Orthopaedics and Trauma*. Available from https://doi.org/10.1016/j.jcot.2019.07.003.

Hassen, H. B., Ayari, N., & Hamdi, B. (2020). A home hospitalization system based on the Internet of things, fog computing and cloud computing. *Informatics in Medicine Unlocked* *20*, 100368.

He, Z., Cai, Z., Yu, J., Wang, X., Sun, Y., & Li, Y. (2017). Cost-efficient strategiesfor restraining rumor spreading in mobile social networks. *IEEE Transactions on Vehicular Technology*, *66*(3), 2789−2800.

Holshue, M. L., DeBolt, C., Lindquist, S., Lofy, K. H., Wiesman, J., Bruce, H., ... Tural, A. (2020). First case of 2019 novel coronavirus in the United States. *The New England Journal of Medicine*, *382*, 929−936. Available from https://doi.org/10.1056/NEJMoa2001191.

Hu, F., Xie, D., & Shen, S. (2013). On the application of the internet of things in the field of medical and health care, Proc. - 2013 IEEE International Conference on Green Computing and Communications and IEEE Internet of Things and IEEE Cyber, Physical and Social Computing. *Green Com-iThings-CPS Com*, 2013, no. August 2013, pp. 2053−2058.

Jamal, A., Narayanasamy, D. D. A. L., Mohd Zaki, N. Q. & Abbas Helmi, R. A. (2019). Large hall temperature monitoring portal, in 2019 IEEE *International Conference on Automatic Control and Intelligent Systems* (I2CACIS), Selangor, Malaysia, pp. 62−67.

Javaid, M., Vaishya, R., Bahl, S., Suman, R., & Vaish, A. (2020). Industry 4.0 technologies and their applications in fighting COVID-19 pandemic. *Diabetes & Mmetabolic Syndrome. Clinical Research & Reviews*. Available from https://doi.org/10.1016/j.dsx.2020.04.032.

Johnson & Johnson. (March 2020). Johnson & Johnson announces a lead vaccine candidate for COVID-19. Available from: https://www.jnj.com/johnson-johnson-announces-a-lead-vaccinecandidate-%25for-COVID-19-landmark-new-partnership-with-u-sdepartment-of-health-human-serv%25ices-and-commitment-to-supplyone-billion-vaccines-worldwide-for-emergency-pan%demic-use.

Khanna, A. & Anand, R. (2016). IoT based smart parking system. In 2016 International Conference on Internet of Things and Applications (IOTA), pp. 266−270. IEEE.

Kim, E., Erdos, G., Huang, S., Kenniston, T. W., Balmert, S. C., Carey, C. D., ... Gambotto, A., Microneedle array delivered recombinant coronavirus vaccines: Immunogenicity and rapid translational development. *EBioMedicine*, *55*, 102743. Available from https://doi.org/10.1016/j.ebiom.2020.102743.

Kumari, R., Kumar, S., Poonia, R. C., Singh, V., Raja, L., Bhatnagar, V., & Agarwal, P. (2020). Analysis and predictions of spread, recovery, and death caused by COVID 19 in India. *Big*

Data Mining and Analytics. Available from https://doi.org/10.26599/BDMA.2020.9020013, IEEE.

Kummitha, R. K. R. (2020). Smart technologies for fighting pandemics: The techno- and human-driven approaches in controlling the virus transmission. *Government Information Quarterly*.

Li, Q., Guan, X., Wu, P., Wang, X., Zhou, L., Tong, Y., et al. (2020). Early transmission dynamics in Wuhan, China, of novel coronavirus-infected pneumonia. *The New England Journal of Medicine*. Available from https://doi.org/10.1056/NEJMoa2001316.

Li, T., Wei, C., Li, W., Hongwei, F., & Shi, J. (2020). Beijing Union Medical College Hospital on "pneumonia of novel coronavirus infection" diagnosis and treatment proposal (V2.0). *Medicine Journal of Peking Union Medical College Hospital*. Available from http://kns. cnki.net/kcms/detail/11.5882.r.20200130.1430.002.html. Accessed 2 Feb 2020.

Liu, T., Hu, J., Kang, M., Lin, L., Zhong, H., Xiao, J., et al. (2020). Transmission dynamics of 2019 novel coronavirus (2019-nCoV). Available from: https://doi.org/10.1101/2020.01.25.919787.

Ma, Y., Diao, B., Lv, X. et al. (2020). Novel coronavirus disease in hemodialysis(HD) patients: Report from one HD center in Wuhan, China. Available from: https://www.medrxiv.org/content/10.1101/2020.02.24.20027201v2. Accessed 14 March 2020.

Medical expert group of Tongji hospital. (2020). Quick guide to the diagnosis and treatment of pneumonia for novel coronavirus infections (third edition). Herald of Medicine. http://kns. cnki.net/kcms/detail/42.1293.r.20200130.1803.002.html. Accessed February, 2020.

Moderna. (March 2020). Moderna's work on a potential vaccine against COVID-19. Available from: https://modernatx.com/modernaswork-potential-vaccine-against-COVID%-19.

Mohammed, M. N., Desyansah, S. F., Al-Zubaidi S., & Yusuf, E. (2020). An internet of things-based smart homes and healthcare monitoring and management system, *Journal of Physics: Conference Series*, 1450, 012079.

Mohammed, M. N., Hazairin, N. A., Syamsudin, H., Al-Zubaidi, S., Sairah, A. K., Mustapha, S., & Yusuf, E. (2020). 2019 novel coronavirus disease (COVID-19): Detection and diagnosis system using IoT based smart glasses. *International Journal of Advanced Science and Technology*, *29*(7s), 954−960.

Mohammed, M. N., Syamsudin, H., Al-Zubaidi, S., Sairah, A. K., Ramli, R., & Yusuf, E. (2020). Novel COVID-19 detection and diagnosis system using IoT based smart helmet. *International Journal of Psychosocial Rehabilitation*, *24*(7).

National Health Commission of People's Republic of China (2020). Prevent guideline of 2019-nCoV. Available from: https://www.nhc.gov.cn/xcs/yqfkdt/202001/bc661e49b5bc487d-ba182f5c49ac445b.shtml. Accessed February 1, 2020.

Park, A. (March 2020). As the first coronavirus vaccine human trials begin, manufacturer is already preparing to scale production to millions. Available from: https://time.com/5807669/coronavirusvaccine-moderna/.

Centers for Disease Control Prevention (CDC) (April 2020). People who are at higher risk for severe illness. Available from: https://www.cdc.gov/coronavirus/2019-ncov/need-extraprecautions/people%-at-higher-risk.html.

Ruktanonchai, N. W., Ruktanonchai, C. W., Floyd, J. R., & Tatem, A. J. (2018). Using Google Location History data to quantify fine-scale human mobility. *International Journal of Health Geographics*, *17*(1), 1−13.

Sardianos, C., Varlamis, I., & Bouras, G. (2018). Extracting user habits from google maps history logs, Proc. 2018 IEEE/ACM *International Conference on Advances in Social Network Analysis and Mining*, ASONAM 2018, pp. 690−697.

Centers for Disease Control Prevention (CDC). (December 2017). SARS Basics Fact Sheet. Available from: https://www.cdc.gov/sars/about/fssars.Html.

Singh, R. P., Javaid, M., Haleem, A., & Suman, R. (2020). Internet of things (IoT) applications to fight against COVID-19 pandemic. *Diabetes & Metabolic Syndrome: Clinical Research & Reviews, 14,* 521−524.

Singh, V., Poonia, R. C., Kumar, S., Dass, P., Agarwal, P., Bhatnagar, V., & Raja, L. (2020). Prediction of COVID-19 coronavirus pandemic based on time series data using support vector machine. *Journal of Discrete Mathematical Sciences & Cryptography.* Available from https://doi.org/10.1080/09720529.2020.1784525.

Singhal, T. (2020). A review of coronavirus disease-2019 (COVID-19). *Indian Journal of Pediatrics, 87*(4), 281−286.

Song, Y., Jiang, J., Wang, X., Yang, D., & Bai, C. (2020). Prospect and application of Internet of Things technology for preventionof SARIs. *Clinical eHealth, 3,* 1−4.

Stoessl, A. J., Bhatia, K. P., & Merello, M. (2020). Movement disorders in the world of COVID-19. Movement Disorders Clinical Practice. In press.

Swayamsiddha, S., & Mohanty, C. (2020). Application of cognitive Internet of Medical Things for COVID-19 pandemic, Diabetes & Metabolic Syndrome: Clinical Research & Reviews.

Vaishya, R., Javaid, M., Khan, I. H., & Haleem, A. (2020). Artificial Intelligence (AI) applications for COVID-19 pandemic. *Diabetes & Metabolic Syndrome. Clinical Research & Reviews.* Available from https://doi.org/10.1016/j.dsx.2020.04.012.

Viola, P., & Jones, M., (2001). Rapid object detection using a boosted cascade of simple Features, Proc. 2001 IEEE Comput. Soc. Conf. Comput. Vis. Pattern Recognition. CVPR 2001, pp. 511−518.

Wang, B., Sun, Y., Duong, T. Q., Nguyen, L. D., & Hanzo, L. (2020). Risk-aware identification of highly suspected COVID-19 cases in social IoT: A joint graph theory and reinforcement learning approach, VOLUME 8, 10.1109/ACCESS.2020.3003750.

Wang, C., & Wang, X. (2020). Prevalence, nosocomial infection and psychological prevention of novel coronavirus infection. *Chinese General Practice and Nursing, 18,* 2−3.

Wang, D., Hu, B., Hu, C., Zhu, F., Liu, X., Zhang, J., . . . Peng, Z. (2020). Clinical characteristics of 138 hospitalized patients with 2019 novel coronavirus_infected pneumonia in Wuhan, China. *Journal of the American Medical Association, 323*(11), 1061.

Wang, M., Cao, R., Zhang, L., Yang, X., Liu, J., Xu, M., . . . Xiao, G. (2020). Remdesivir and chloroquine effectively inhibit the recently emerged novel coronavirus (2019-nCoV) in vitro. *Cell Research, 30*(3), 269−271.

World Health Organization (2020a). Laboratory testing for 2019 novel coronavirus (2019-nCoV) in suspected human cases, vol. 2019, no. January, pp. 1−7.

World Health Organization (2020b). Novel coronavirus (2019-nCoV) advice for the public. Available from: https://www.who.int/emergencies/diseases/novel-coronavirus-2019/advice-for-public.

WHO (April 2020c). *Coronavirus Disease 2019 (COVID-19) SituationReport 87.* Available from: https://www.who.int/docs/defaultsource/coronaviruse/situation-reports/%20200416-sitrep-87-COVID-19.pdf?sfvrsn = 9523115a_2.

WHO. (April 2020d). *Coronavirus Disease 2019 (COVID-19) SituationReport 79.* Available from: https://www.who.int/docs/defaultsource/coronaviruse/situation-reports/%20200408-sitrep-79-COVID-19.pdf?sfvrsn = 4796b143_6.

WHO Emergency Dashboard. Available from: https://COVID19.who.int/? gclid = CjwKCAjwx9_4BRAHEiwApAt0zh1NF5dqq002bZYmIVgSIRygPi1dT1wVLtsYzY379-f3eti8SPK4T-xoCbbgQAvD_BwE.

Yang, T., Gentile, M., Shen, C. F., & Cheng, C. M. (2020). Combining point-of-care diagnostics and internet of medical things (IoMT) to combat the COVID-19 pandemic. *Diagnostics*. Available from https://doi.org/10.3390/diagnostics10040224.

Zamani, N. S., Mohammed, M. N., & Al-Zubaidi, S. (2020). Design and development of portable digital microscope platform using IoT tTechnology, in IEEE International Colloquium on Signal Processing & Its Applications (CSPA 2020).

Zheng, S. Q., Yang, L., Zhou, P. X., Li, H. B., Liu, F., & Zhao, R. S. (2020). Recommendations andguidance for providing pharmaceutical care services during COVID-19 pandemic: A China perspective. *Research in Social and Administrative Pharmacy*.

Zhou, P., Yang, X. L., Wang, X. G., Hu, B., Zhang, L., Zhang, W., et al. (2020) Discovery of a novel coronavirus associated with the recent pneumonia outbreak in humans and its potential bat origin. bioRxiv. Available from: https://doi.org/10.1101/2020.01.22.914952.

Chapter 3

Role of mobile health in the situation of COVID-19 pandemics: pros and cons

Priyanka Bhaskar[1] and Sunita Rao[2]
[1]Department of Management Studies, Swami Keshvanand Institute of Technology, Management & Gramothan (SKIT), Jaipur, Rajasthan, India, [2]School of Engineering and Technology, Jaipur National University, Jaipur, Rajasthan, India

3.1 Introduction

The impact of the disease caused by a novel coronavirus (2019-nCoV) generalized as COVID-19 is affecting each nuke and corner of the world with more or less similar complications. The disease is reported to affect more than 4 million individuals, tainted by the infection so that COVID-19 limitations are applied to the everyday life of people worldwide (Worldometer, 2020). The measures taken to decrease the spread of novel coronavirus or forestall contamination followed the cleanliness with sanitization rules (Chen et al., 2020). The most significant precaution is washing hands many times a day to slow down the spread of the infection at the social orders with maintaining social distancing (Nakada & Urban, 2020). There was a global level of lockdown that has been witnessed. Researchers have declared that lockdown is a temporary solution to limit the spread of infection in developing and developed nations equally (Tobías et al., 2020). In support of the statement, approximately 214 nations announced the total number of affirmed COVID-19 cases in their country daily (Chakraborty & Maity, 2020). Governments have taken severe limitations to be followed by the population, such as declared holidays for schools, work from home culture for offices, quarantine period for the containment zones those has high number of cases, and lockdown to stop the COVID-19 flare-up. The lockdown days were with variable in different nations decided by the local authorities to limit the COVID-19 impact on the population. A few nations have increased the lockdown time-period by numerous days due to the COVID-19 effect on general society. Some of the governments continued the lockdown until this report.

Cyber-Physical Systems. DOI: https://doi.org/10.1016/B978-0-12-824557-6.00005-4

However, the last day of lockdown in many countries was witnessed as on May 5, 2020, with a remarkable example given by Ireland, with a curfew for 68 days, which has the most extended lockdown period during which a total of 21,000 COVID-19 cases were approved till May 5, 2020. Due to the highest number of cases in Spain, it has been imposed with a lockdown for 53 days (Chen et al., 2020).

During the lockdown period, a severe impact has been witnessed on all the sectors along with the healthcare delivery system, as the general public has been advised to stay indoors and not to visit any social place including clinics, hospitals, dispensaries, public healthcare centers, community health centers, and so on (Kasthuri, 2018). In this context, the patients who were planned with surgeries were faced with problems and were in a queue of routine check-ups. To take a way out of this, several online platforms, by the aid of software, came up with the solution for virtual patient − doctor meetings as a part of the popular term mobile health, which can be a future way to improve the value, proficiency, viability, and responsiveness of the healthcare frameworks to be ready for emergency management, remote consultation etc. as shown in Fig. 3.1.

The role played by mobile centers ought to give the proof to legitimize arrangements that will empower the ideal incorporation of mobile facilities into any mobile health (mHealth) care delivery framework (Attipoe-Dorcoo, 2018). With national endeavors for social determinants of good health, there is presently, like never before, the need to consider firm government policies

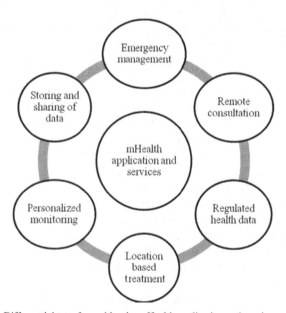

FIGURE 3.1 Different jobs performed by the mHealth application and services.

for mobile healthcare facilities. Even though there are current financial hindrances faced by the population of developing nations tending to well-being, weak patient care results will likewise assume a job in adding significant worth-based installment models (Malone et al., 2020). There are impermanent administrative exemptions that have been made by the Centers for Medicare and Medicaid to repay the utilization of apartments as emergency clinics to help with the attempt to battle the COVID-19 pandemic (Muchmore, 2020).

3.2   Implementation of a training module for the mHealth care worker

Effective utilization of any new technology is only possible when there is a critical learning process at ground level. mHealth is a new scientific innovation for many developing nations, in terms of medical curricula, and there is a lack of adequate understanding about when and how it works (Moore et al., 2017). It has been observed that there is a severe shortage of qualified technical skilled individuals in both developed and developing nations for mHealth. They are facing considerable challenges in identifying and developing technical skills to those individuals. American Medical Informatics Association took the initiative by launching suitable graduation programs like short courses, a project-centric model to provide healthcare practitioners training specifically for developing countries like India, Africa, and Latin America (Hersh, 2008).

mHealth requires implementing academic models with appropriate teaching structure in digital communication skills, deployment of telehealth products and services, and knowledge of the regulatory and legal issues (Slovensky, Malvey, & Neigel, 2017). Even educational institutions presume that clinicians should learn mHealth skills as a part of their regular curriculum. Current medical graduates are considered as "digital natives," the reason being they are familiar with technology and could implement skills into their professional practices to contact and examine patients (Malvey & Slovensky, 2014). The generation can be considered the first generation of medical graduates ideal for incorporating innovations in health care. Examples have been set by the advanced degree program, launched with a partnership with the local universities and its healthcare delivery system, funded by Bill & Melinda Gates Foundation https://www.amia.org/GPP (https://www.amia.org/education/programs-and-courses).   Technological expertise improves healthcare efficiency, safety, and reduced costs (Hersh, Margolis, Quirós, & Otero, 2008).

The practical exercises for skill enhancement programs include developing health informatics applications, advancing wellness computing tools, and data science. The research examined the association between specific practical skills training and user's mindset in adopting technology (Sapci, 2019).

E-platform like mHealth apps have a broader scope than in-person healthcare management as it has a greater reach among the users, and dependency on physical travel is nullified. The Accreditation Council for Graduate Medical Education organized several programs that include seminars, case-based learning, teaching, administration, assessment, and upgrading quality in health management to par with conventional health diagnostic. A significant paradigm shift is required in building a productive e-platform in the health-care system through significant improvement in the technological skills of both medical professionals and the patients. The technological skill helps to handle the mHealth and Smartphone apps help to achieve the goals of pro-viding healthcare at doorstep (Hilty, Chan, Torous, Luo, & Boland, 2019). While designing the program, if the algorithm initially provides the terrible results, it will have adverse effects on health-related outcomes and user engagement. To avoid adverse effects and harmful results, it is better to con-sult a mobile device specialist (Guha, 2018).

3.3 Government policies for the scale-up of the mHealth services

As mobile clinics can be significantly chosen for the service conveyance, particularly after a debacle has caused fixed offices to close, this care model has not been generally supported earlier (Rassekh, Shu, Santosham, Burnham, & Doocy, 2014). It has been proposed by three general ways to deal with the upgradation of mobile clinics programs and their framework in complete combination. The first place has to be perceived by the financial commitment for the mobile center projects meant for the human healthcare services. The mobile clinics offer some benefit to the general population residing in the rural and urban regions by offering a network as far as fore-stalled visits to the emergency services. To evaluate such advantages, there is significant work to be finished (Hill et al., 2014). In connection it is important to investigate mobile clinics' economic effects in terms of the three checkpoints: first of all the decreased per capita normal expenses of care, advantages to populace well-being, and upgrades intolerant fulfillment. Through this perspective, we will have the option to work with policy-makers, suppliers, and payers to characterize suitable repayment plans for mobile facility program administrations.

Second, it has been accepted that a particular government financing proj-ect ought to be designed to give required financial support that will permit both the development and extension of the number of versatile center pro-jects. This will not only guarantee that the facilities are promptly fused into the current medicinal services framework, but also crisis readiness. A case of how to make such an arrangement realistic is by considering versatile centers on the administration models proposed in the national COVID-19 observa-tion framework (McClellan, Gottlieb, Mostashari, Rivers, & Silvis, 2020).

These models are proposed to expand the limit of treating patients in detachment offices, incorporating portable well-being facilities through disease, and recuperating patients. These endeavors will be fruitful with improved government repayment models that spread network-based assets, just as state what is more, neighborhood well-being coordination (McClellan et al., 2020). Setting up these frameworks presently, as shown in Fig. 3.2, will likewise fortify general well-being and human services readiness for future flareups on the different levels, which can be a government, public initiation, or a private company-owned business model based on symptoms of tracking, information sharing, or follow-up basis.

Third, it has been proposed that making national subsidy schemes on the public − private partnership model can help to give support to the existing model to serve better. This will permit versatile center projects to set up close shared contributions with different partners in the human services framework. For instance, repayment for mHealth is innovation in versatile facilities, and the capacity to allude and explore patients in a far-reaching constant way. This methodology, in blend with Geographic Information Systems (GIS)-based course streamlining calculations, could be utilized to decide the need for territories, particularly country zones where fixed offices are shutting at quick rates (Attipoe-Dorcoo, 2018). In addition, the turn of events of online applications is dependent on information gathered through the mHealth Map, such as in the case of the program of Harvard Medical

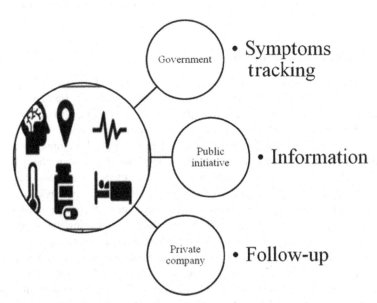

FIGURE 3.2 mHealth care management at three different modes to serve the population. *mHealth*, Mobile health.

School's Family Van, which has been utilized by state authorities to guide versatile facility assets to divert toward zones of high need. Existing geographic calculations could be utilized to decide such areas of need, and the versatile facilities could be utilized to arrive at populaces effectively. Moreover, existing proportions of the more extensive scope of network need could help direct portable facilities to networks needing anticipation administrations, just as address the issues of imbalance experienced in numerous underresourced networks (Farmer, 2020).

3.4 Popular models of mHealth serving for pandemic COVID-19

The World Health Organization characterizes mHealth as: "medical and public health practice supported by mobile devices, such as mobile phones, patient monitoring devices, personal digital assistants and other wireless devices." The developing prominence of cell phones in social insurance has been offered to the Mobile Health, or mHealth, industry during COVID-19, as reported in Table 3.1. Fueled by advancement and expanding networks worldwide, mHealth innovation is developing at an incredible rate. For instance, in 2018, there were less than approx. 3 lakh health-related applications accessible for download, practically twofold the number of applications on the top application stores in 2015. It is considered that the versatile human service innovation advertises getting a charge out of enormous achievement and is anticipated to arrive at an estimation of €53 billion by 2020 (Franco, 2015).

3.5 Ethical consideration

Due to various reasons, mHealth became more popular in the past few years, spreading as a practice working style for doctor-to-doctor frameworks in numerous countries worldwide over all asset levels (Blom, 2018; Thomas, 2018). However, it presents specific moral concerns (Sharp & O'Sullivan, 2017). Concerning media data used for interactions, clinical pictures are communicated or distributed without having either the clinician or the patient utterly mindful of the break-in only security glitch that may happen (McCartney, 2018). In addition, the utilization of clinical pictures coupled with the online meeting is equivalent to telemedicine, which is continuously developed and designed. There are works in shields for securing the dispersion and capacity of the clinical records that online networking does not give.

Further, sometimes clinical pictures at this point exclusively, if not taken by proficient professionals at the same time, may need explicit capability (Palacios-Gonzalez, 2015; Sharp and O'Sullivan, 2017). Suggestions to alleviate concerns relating to the moral standards of self-rule, well-being, and equity are accessible in analytical writing and proficient rules, such as medication, nursing,

TABLE 3.1 Most popular web-based applications for COVID-19.

Geographical location	Application	Features
Singapore	https://mothership.sg/2020/03/trace-together-COVID/	It does not collect any other personal data or location data (GPS, Wi-Fi, cellular networks) but is used for tracing.
Israel	https://medium.com/@oleiba/hamagen-fight-coronavirus-and-preserve-privacy-b1631693bb46	It detects a sick person's location. Notifies the user and provides him with the updated instructions.
Austria	https://participate.roteskreuz.at/stopp-corona/	The infected person triggers the warning of close contacts after medical confirmation of infection past 54 h.
Bulgaria	https://coronavirus.bg/arcgis/apps/opsdashboard/index.html#/ecacd239ee7e4fba956f7948f586af93	Real-time link between citizen's authorities and the healthcare system.
Iceland	https://www.covid.is/app/en	Contract tracing app "Ranking C-19" helps to analyze individuals' travel and trace their movements.
Cyprus	http://covid-19.rise.org.cy/	An application named TRACER for racing of location.
Czech Republic	github.com/covid19cz/erouska-androidandgithub.com/covid19cz/erouska-ios	Bluetooth proximity tracing app, created by the covid19cz, endorsed by the Ministry of Health. The app uses pseudonymous IDs; the phone's data are provided to the central system only with explicit consent.
France	https://www.servicepublic.fr/particuliers/actualites/A13927%20follow-up%20of%20COVID-19	Software that analyzes the COVID-19 symptoms. Medical survey is sent to patients every day

(Continued)

TABLE 3.1 (Continued)

Geographical location	Application	Features
Germany	https://experience.arcgis.com/experience/478220a4c454480e823b17327b2bf1d4	
Ireland	https://www.hsecovid19.ie/	Data sharing of COVID-19 health information of a patient during self-isolation.
Italy	http://www.regione.lazio.it/rl/coronavirus/scarica-app/ https://www.dedalus.eu/dedalus-per-covid-19/	Telehealth and monitoring with a general practitioner in Lazio region, screening, active surveillance at home, tutorial, chatbot, telehealth.
India	https://www.aarogyasetu.gov.in/	The contract tracing app helps to analyze travel and trace the movements of COVID patients.
The Netherlands	https://apps.apple.com/nl/app/covid-19-medisch-dossier/id1502322865	Public information app with Q&A, phone numbers, and map of the Netherlands locating infections.
Norway	https://helsenorge.no/coronavirushttps://www.simula.no/news/eng-simula-working-Norwegian-institute-public-health	To monitor the spread of the infection and to assess the effects of the infection control measures.
Poland	https://www.gov.pl/web/koronawirus/wykaz-zarazen-koronawirusem-sars-cov-2	Contact tracing and self-diagnostic app.
Portugal	https://covid19.min-saude.pt/ or https://www.sns24.gov.pt/	Tracing app but on promotion, monitoring, and self-care.
Spain	https://covid19.es/	Chatbot using AI and Natural Language named Hispabot-Covid19

(Continued)

TABLE 3.1 (Continued)

Geographical location	Application	Features
Brazil	https://play.google.com/store/apps/details?id = br.gov.datasus.guardioes&hl = en	Symptoms, how to prevent, what to do in case of suspicion and infection.
Vietnam	https://www.opengovasia.com/vietnam-launches-health-app-to-manage-COVID-19/	Citizens can update their daily health status and provide information.

or designing (Albrecht & Fangerau, 2015; Carter, Liddle, & Hall, 2015). The piled up data are disaggregated only from time to time as per the phases of mHealth applications (advancement, assessment of adequacy, execution, and scale-up). Their acknowledgment over a scope of partners is unsure; their low-asset settings have gotten constrained consideration (Cresswell, Bates, & Sheikh, 2013). The examination expects to fill those information holes by researching what partners from various settings and foundations concur and organize to handle self-sufficiency, security, and equity concerns. Issues around unapproved or vindictive access to sensitive information were much considered in mHealth's ethics, especially in current circumstances (Kotz, Gunter, Kumar, & Weiner, 2016). In any event, when well-being information is put away on HIPAA (Health Insurance Portability and Accountability Act of 1996)-compliant workers, essential prominent hacks uncover that the data are as yet powerless against bargain (Peterson, 2016). The information gathered by mHealth applications might be weaker still, as the versatile sensor or nearby information stockpiling is typically worn or conveyed by the patient or subject. Like regular stockpiling of human services information, mHealth innovation ordinarily depends on cloud figuring, and rehashed hacks have exhibited that cloud stockpiling is a long way from being resistant to security penetration. Dissimilar to ordinary stockpiling of medicinal services information, mHealth innovation necessitates that conceivably delicate well-being data are gathered or put away at the gadget level. For instance, patients not utilizing mHealth innovation may have psychological sickness insights in their clinical records, especially during pandemics.

Patients who are well-versed of use of mHealth innovation to screen their psychological sickness indications will at present have the judgments and treatment plans put away on a data cloud. However, now they will likewise have new classifications of delicate and individual data on their cell phones, for instance, and this presents another degree of hazard. The guarantee of expanding availability to social insurance meets a few more difficulties. For instance, mHealth intercessions like the ones depicted above, utilizing instant messages, may not reach the individual need care, either in light of the fact those are ignorant, or neighborhood dialects are not bolstered by versatile phones (Bullen, 2013).

On the other hand, applications require cell phones with a quick web association and capacity to connect with these telephones, in this manner barring specific gatherings (e.g., low-pay gatherings or older individuals with less expertise with cell phones). These issues bring up the confusion of whether the individuals who need better consideration or improved access to mind at present are being served by mHealth advancements. These angles undermine the guarantees of availability to healthcare services and raise issues whether mHealth intensifies instead of mitigating social equity issues.

3.6 Superiority of mHealth services over other available services

mHealth devices and applications help in therapeutic progress compared to conventional treatment due to the diagnosis of illnesses. Documentation/billing is perceived as an unregulated burden generated by an increased number of patients in hospitals. The mHealth applications and their usage reduce patients' movement to the hospitals and help medical representatives document their disease treatments through a simple yet effective structure (Zeman, Moon, McMahon, & Holley, 2018). Millions of medical representatives supported the growing technological developments in health care, and the implementation of voice-interactive tools is one of them in the everyday lives in patient's treatment. Voice-interactive tools are anticipated as a long-term practical solution in home documentation and delivery of information, which helps in saving time and extra costs. The importance of health information outside the hospital generated at home is a big resolute to the medico personnel. On occasions, the patient's admission and discharge is a big challenge for clinical supervision. The recommendation of voice-interactive tools not only improves record maintenance but also helps in communicating with different patients, care providers, and caregivers as technological solutions in telemedicine setup (Sezgin et al., 2020).

Rigid contamination control of transmissible diseases such as COVID-19 is a basic necessity at treatment centers, among the specialized isolating amenities where patient's treatment has been done. Major obstacles have been faced previously, was poor documentation and examination of patient's record as seen case of Ebola pandemic in West Africa during the year 2014−16, where paper-based information at Ebola treatment centers was a proven failure in the documentation of patient's information in terms of volume, quality, and secrecy. The electronic mHealth framework can tackle these challenges, including implications for medical treatment, monitoring, and analysis. A software named as OpenMRS Ebola was used for tracking patient's information (enrollment, bed allocation, and discharge); documentation of critical signs and symptoms; scheduling and evaluation of prescription; test results; doctor's observations and facts in contagious areas (Oza et al., 2017). mHealth apps have recently accelerated job allocation techniques with patient care, particularly in low and middle-income countries. Fewer studies address user-friendliness and practicable difficulties that might affect the future direction, and few were implemented for noncommunicable diseases such as hypertension. In Western Kenya, the DESIRE (Decision Support and Integrated Record-keeping) tool was developed to test the performance of its repetitive functionality to support rural clinicians in treating hypertension cases at the community level with limited resources (Vedanthan et al., 2015). In other cases of diseases, which requires maintenance of data about the lifelong history of patients, as in the case of thalassemia (genetic blood disorder), many challenges have been faced by Malaysia's public health management personnel due to lack of technological skills by the users to monitor the frequent monitoring checks that the patient is supposed to undergo. During their lifetime, regular supervision of patients needs modern technological development to improve the efficiency and effectiveness to achieve success in thalassemia treatment. mHealth where smartphone plays an assistive role because of its ubiquity and accessibility as a personal device that can support users (patients, care professionals, and clinical) in treating thalassemia disease and position as a way that comes between patient and doctors in health management (Bal et al., 2014).

Therefore, it is essential to have a convenient yet effective patient health-related data capturing mechanism for subsequent utilization in clinical analysis of diseases and tracking parameters for observing trends and patient health changes. Documentation of treatment is thus the pivot point for the successful evolution of mHealth application as it proves to be cost-effective and less time-consuming utility as a medium of interaction between the patient and the medical personnel.

3.7 Probability of conflict of interest between user and service provider

Multifold utilization of mHealth applications is possible when in different health and fitness monitoring areas, the evolution of the mHealth applications occurs with equal consideration of both the end-user and the application developers or creators. At the interface, it seems a visible and transparent domain; however, considering the magnitude of data, there can be implications on developing a business ambition around the quantum of information through unverified and, at times, unethical channels or mechanisms that help to multiply revenue-generating sources.

More often, most of the mHealth applications available on multiple platforms have a potential to influence the user behavior by increasingly compelling ways as these applications not only influence the health patterns of patients but also the economic behaviors emerging from the culmination of both health and commercial content in ways those are significantly challenging to detect and have a trace upon. Hence user autonomy can be impacted by unfair commercial and ethical practices. Therefore it is essential to evaluate the user expectations from mHealth applications regarding possible benefits from the applications' usage, hence enabling mHealth applications a preferred consulting channel for the global mass. The patient's expectations are as per

1. Conscious information: thoughts, perceptions, beliefs, etc.;
2. Preconscious information: memory, past experiences regarding a particular illness and treatment;
3. Unconscious information: fears, selfish motives, instincts, etc.

The above factors make every patient distinctively unique, and therefore each one needs a customized resolution through a robust diagnostic experience (Boag, 2017). Patients' underlying expectation is to have a cost-effective, simple to use, and secure platform in cyber-physical space. From the developer perspective, the commercial aspect regarding the number of application downloads, review ratings, and extraction of users' relevant data for further analytics are the key imperatives.

3.8 Legal consideration

The legal consideration of data gathered during the diagnostic interaction between the patient and the professional is also a significant point of concern for most of the people who had chosen to interact on the mHealth platforms globally. Though the patient's health profile becomes accessible for the health professionals by one click away, it has some aspects of unwanted usage and data exploitation. More often than not, the interaction platforms used by either of the parties, that is, the patient or the medical personnel, have parallel access to multifold social networking zones, which at times might lead to the leakage of sensitive patient-generated health data in the social networks of either of the third parties in the data exchange.

To analyze the concern to understand the ownership of data at various levels of transit in the digital space by the elementary or so to say notional "Terms of Service" lays the foundation stone of data ownership across most of the social media channels and mostly has a well-defined statement of data ownership at the onset of an application before its full-fledged usage, predominantly in an mHealth scenario. As a preventative practice, the medical representatives or the doctors should first educate the patients for the mHealth apps at the very beginning, and warning signs for limitation of these apps should be flashed as an initial procedure (Yang & Silverman, 2014).

In general, these terms and conditions decide the usage of data in diagnosis. Hence, it is necessary to go through the same before the intended usage for patient-generated health data. Once a data repository is formed, at the mercy of the site or application owner, there is concern arises about how and when to use it. However, there are still specific provisions for the users in countries like the United States under the Health Insurance Portability and Accountability Act of 1996 (Petersen & DeMuro, 2015).

In most diagnosis cases through the mHealth platforms, professionals' treatment is conducted knowingly and unknowingly with cognitive bias. The reliability of data captured by the patient is also a gray area as far as accuracy is concerned. The patient-generated health data are mostly recorded by the patient, and there are chances of being erroneous data due to two conditions:

1. The physical well-being of the patient might not support him/her while recording the data.
2. Lack of skill or training to capture health data by the patient.

The above two constraints might well lead to data transfer with specific errors to the medical representative at the diagnostic end of the mHealth platforms. This might lead to a failure of judgment in the future course of medical action with the patient. Therefore it becomes undeniable that the patient's self-monitoring health app has to be judiciously working. It decides the subsequent treatment actions for the doctors or health professionals in the subject. Therefore in such cases where if one has to define the ownership of the medical personnel in terms of erroneous diagnosis basis, the inaccurate information provided by the patient becomes an area of legal intervention if the need arises. But, unfortunately, the on-date legal framework does not conclude much about the resolution of such scenarios where there is a discrepancy in data at the point of origin and the point of application.

3.9   Protection of privacy of end-users

According to HIPAA regulation Electronic Protected Health Information (ePHI) consists of 18 different demographic parameters to recognize a user. The creation of ePHI aims to store, receive, or transfer health information electronically, keeping in mind the security rules or standards set by HIPAA. Confidentiality ensures that without adequate patient authorizations, ePHI is not disclosed. Integrity ensures that ePHI is accessed by relevant and approved parties who come under the healthcare organizations. Availability permits patients' ePHI only if they follow HIPAA security standards. Examples of ePHI are name, address, email address, medical records, etc.

As mHealth applications' usage erupts in the global landscape, the need for assuring no pilferage of information gathered directly or indirectly during the interaction is of supreme importance. The nature of diagnostic requires a close interaction in the cyber-physical space and therefore has a potential risk exposure of valuable information to the domains unintended for future hazards. The potential risk can be bifurcated in privacy-related and security-related domains. Therefore various techniques and tools for assessing security and/or privacy need to be closely assessed and evaluated for mitigating risk and creating a robust platform for interaction (Nurgalieva, O'Callaghan, & Doherty, 2020).

The majority of applications developed globally are based on two platforms: Android or iOS, depending on the end interface available to the consumer and the service provider, thereby enabling an extensive data transfer back and forth between both the medical professionals and patients. Therefore the reliability of secure third-party servers and trustworthy internet communication services is a critical cog in the development wheel of the mHealth applications in terms of customer experience and effective treatment (He, Naveed, Gunter, & Nahrstedt, 2014). The significant transition of the physical way of diagnosing ailment to a diagnostic setup in a virtual environment would also need a safe and secure payment structure with a hassle-free experience for both the professionals and the patients; henceforth the need for secured payment gateways and online financial transactional support with monitoring and governing authority is also a must for the mushrooming of this setup.

Owing to the global upsurge usage of mHealth applications for diagnostics and treatments through a seamless virtual experience digitally, a well-defined sequence of procedures can be utilized as a ready reference for both parties involved for substantial benefit. It can be an essential document for problem resolution dealing with data privacy, security, and confidentiality concerns. Moreover, a precise cut formulation of guidelines would also enforce trust in the interaction, and this will further foster the bond between the professionals involved with the patients. On analyzing the various source points of data security, privacy, and confidentiality, one can conclude that there is a significant amount of overlap in the mentioned three areas. Hence, reasonable documentation is, therefore, also required to distinguish between the three (Bhatnagar et al., 2020; Singh et al., 2020; Spigel, Wambugu, & Villella, 2018).

3.10 Conclusion

There is a requirement for both creative- and strategy-based answers for pandemic and future healthcare delivery systems. mHealth care facilities are a crucial piece of the arrangements, and the present situation demands to perceive the more extensive potential for current and future healthcare emergencies. mHealth apps are technologically new and continuously improving day by day. It is necessary for graduate education and training to begin with the ethical and legal utilization of mHealth applications. In the present scenario of COVID-19, there is a requirement of constant and relevant training in mHealth apps. Training and supervision hold the key to scaling up of mHealth applications globally as it is continually improving process and the effort to be put up initially if very high and regular practice for usage at both supervision and patient-level is made.

3.11 Future prospects

Spread of several different communicable diseases has been witnessed every few years; overuse of medical drugs is one of the reasons for outbreak of pandemics such as COVID-19. A good networking and sharing of authentic information over a global healthcare system can provide a solution for early detection of such problems. If symptoms of every novel health problem are shared at the onset of disease, it can help the healthcare professionals to solve the problem on time. For this a well-developed health-care delivery system for emergency situations can save a lot of lives as well as money. The education system must include a well-designed mHealth curriculum as a mandatory course, along which training should be provided in terms of ethical and legal utilization of mHealth applications. Also, there is a requirement of constant and relevant training for mHealth apps for essential integrity, patient privacy, and proper arrangements as per the geographical locations. Health professionals should review the existing literature regarding the guidelines and support of the application for mHealth care. Clinically the health professionals must indulge in learning and discussion with those practitioners who are already practicing the mHealth apps in the medical care system. Clinical approval is required about the effectiveness, authentication, and usefulness of the mHealth apps for specific users at the community level. Experiences learned from previous crises strongly recommend the mHealth apps, which are well-designed, user friendly, and pretested by the clinicians for future needs.

References

Albrecht, U. V., & Fangerau, H. (2015). Do ethics need to be adapted to mHealth? *Studies in Health Technology and Informatics, 213*, 219−222.

Attipoe-Dorcoo S. (2018). *An overview of costs, utilization, geographical distribution & influence of mobile clinics in rural healthcare delivery in the United States.* Diss. The University of Texas School of Public Health.

Bal P., Shamsir S., Warid N., Yahya A., Yunus J., Supriyanto E., & Ngim C. F. (2014). mHealth application: Mobile thalassemia patient management application. In *2014 IEEE Conference on Biomedical Engineering and Sciences*, 792−796.

Bhatnagar, V., Poonia, R. C., Nagar, P., Kumar, S., Singh, V., Raja, L., & Dass, P. (2020). Descriptive analysis of COVID-19 patients in the context of India. *Journal of Interdisciplinary Mathematics*, 1−16.

Blom L. (2018). *mHealth for image-based diagnostics of acute burns in resource-poor settings: Studies on the role of experts and the accuracy of their assessments*; https://openarchive.ki.se/xmlui/handle/10616/46382.

Boag, S. (2017). *Conscious, preconscious, and unconscious* Springer Nature *Encyclopedia of Personality and Individual Differences* (pp. 1−8). Springer.

Bullen, P. (2013). Operational challenges in the Cambodian mHealth revolution. *Journal of Mobile Technology in Medicine*, 2, 20−23.

Carter, A., Liddle, J., Hall, W., et al. (2015). Mobile phones in research and treatment: Ethical guidelines and future directions. *JMIR mHealth uHealth*, 3, e95.

Chakraborty, I., & Maity, P. (2020). COVID-19 outbreak: Migration, effects on society, global environment and prevention. *Science of the Total Environment*, 138882. Available from https://doi.org/10.1016/j.scitotenv.2020.138882, 138882.

Chen B., Liang H., Yuan X., Hu Y., Xu M., Zhao Y., ...Zhu X. (2020) Roles of meteorological conditions in COVID-19 transmission on a worldwide scale, *MedRxiv*, https://doi.org/10.1101/2020.03.16.20037168.

Cresswell, K. M., Bates, D. W., & Sheikh, A. (2013). Ten key considerations for the successful implementation and adoption of large-scale health information technology. *Journal of the American Medical Informatics Association: JAMIA*, 20, e9−e13.

Farmer B. (2020). *Nashville Public Radio. Long-standing racial and income disparities seen creeping into COVID-19*. Available at: https://khn.org/news/covid-19-treatment-racial-income-healthdisparities Accessed 06.04.20.

Franco J. (2015). *mHealth market by devices: Global opportunity analysis and industry forecast, 2014−2020: Allied Market Research*; Available: https://www.alliedmarketresearch.com/mobile-health-market.

Guha M., (2018). *Mobile health: Sensors, analytic methods, and applications.*

He D., Naveed M., Gunter C.A., & Nahrstedt K. (2014). Security concerns in Android mHealth apps. In *AMIA Annual Symposium Proceedings*, vol. (2014), 645.

Hersh, W. (2008). Health and biomedical informatics: Opportunities and challenges for a twenty-first century profession and its education. *Yearbook of Medical Informatics*, 17(01), 157−164.

Hersh, W., Margolis, W. A., Quirós, F., & Otero, P. (2008). Building a health informatics workforce in developing countries. *Health Affairs*, 29(2), 274−277.

Hill, C. F., Powers, B. W., Jain, S. H., Bennet, J., Vavasis, A., & Oriol, N. E. (2014). Mobile health clinics in the era of reform. *The American Journal of Managed Care*, 20(3), 261−264.

Hilty, D. M., Chan, S., Torous, J., Luo, J., & Boland, R. J. (2019). A telehealth framework for mobile health, smartphones, and apps: Competencies, training, and faculty development. *Journal of Technology in Behavioral Science*, 4(2), 106−123.

Kasthuri, A. (2018). Challenges to healthcare in India—The five A's.". *Indian Journal of Community Medicine: Official Publication of Indian Association of Preventive & Social Medicine*, 43(3), 141−143. Available from https://doi.org/10.4103/ijcm.IJCM_194_18.

Kotz, D., Gunter, C. A., Kumar, S., & Weiner, J. P. (2016). Privacy and security in mobile health: A research agenda. *IEEE Computer Society*, 49(6), 22−30. Available from https://doi.org/10.1109/MC(2016);185.

Malone, N. C., Williams, M. M., Fawzi, M. C., Bennet, J., Hill, C., Katz, J. N., & Oriol, N. E. (2020). Mobile health clinics in the United States. *International Journal for Equity in Health*, 19(20). Available from https://doi.org/10.1186/s12939-020-1135-7.

Malvey, D. M., & Slovensky, D. J. (2014). *mHealth: Transforming healthcare*. Springer.

McCartney, M. (2018). Margaret McCartney: If you don't pay for it you are the product. *British Medical Journal (Clinical Research ed.)*, *362*, k3249.

McClellan M., Gottlieb S., Mostashari F., Rivers C., & Silvis L. (2020). *A national COVID-19 surveillance system: Achieving containment.* Available at: https://healthpolicy.duke.edu/sites/default/files/atoms/files/covid-19_surveillance_roadmap_final.pdf Accessed 13.04.20.

Moore, M. A., Coffman, M., Jetty, A., Klink, K., Petterson, S., & Bazemore, A. (2017). Family physicians report considerable interest in, but limited use of, telehealth services. *The Journal of the American Board of Family Medicine*, *30*(3), 320−330.

Muchmore S. (2020). *Dorm rooms as hospitals, ER telehealth: CMS creates 'Unprecedented' flexibility as COVID-19 rages on,* Available at: https://www.healthcaredive.com/news/dorm-rooms-ashospitals-er-telehealth-cms-creates-unprecedentedflexibi/575174/? MessageRunDetailID = 1573199395&PostID = 13227449.

Nakada, L. Y. K., & Urban, R. C. (2020). COVID-19 pandemic: Impacts on the air quality during the partial lockdown in São Paulo state, Brazil. *The Science of the Total Environment*, *730*. Available from https://doi.org/10.1016/j.scitotenv.2020.139087.

Nurgalieva, L., O'Callaghan, D., & Doherty, G. (2020). Security and privacy of mHealth applications: A scoping review. *IEEE Access*, *8*, 104247−104268.

Oza, S., Jazayeri, D., Teich, J. M., Ball, E., Nankubuge, P. A., Rwebembera, J., & Walton, D. (2017). Development and deployment of the OpenMRS-Ebola electronic health record system for an Ebola treatment center in Sierra Leone. *Journal of Medical Internet Research*, *19*(8), e294.

Palacios-Gonzalez, C. (2015). The ethics of clinical photography and social media. *Medicine, Health Care, and Philosophy*, *18*, 63−70.

Petersen, C., & DeMuro, P. (2015). Legal and regulatory considerations associated with use of patient-generated health data from social media and mobile health (mHealth) devices. *Applied Clinical Informatics*, *6*(1), 16. Available from https://doi.org/10.4338/ACI-2014-09-R-0082.

Peterson A. (2016). *Why hackers are going after healthcare providers,* The Washington Post (March 28, 2016). https://www.washingtonpost.com/news/the-switch/wp/2016/03/28/why-hackers-are-going-after-health-care-providers/?utm_term = .dc4883252a7d.

Rassekh, B. M., Shu, W., Santosham, M., Burnham, G., & Doocy, S. (2014). An evaluation of public, private, and mobile health clinic usage for children under age 5 in Aceh after the tsunami: Implications for future disasters. *Health Psychology & Behavioral Medicine*, *2*(1), 359−378.

Sapci, A. H., & Sapci, H. A. (2019). Digital continuous healthcare and disruptive medical technologies: m-Health and telemedicine skills training for data-driven healthcare. *Journal of Telemedicine and Telecare*, *25*(10), 623−635.

Sezgin, E., Noritz, G., Elek, A., Conkol, K., Rust, S., Bailey, M., & Huang, Y. (2020). Capturing at-home health and care information for children with medical complexity using voice interactive technologies: Multi-stakeholder viewpoint. *Journal of Medical Internet Research*, *22*(2), e14202.

Sharp, M., & O'Sullivan, D. (2017). Mobile medical apps and mHealth devices: A framework to build medical apps and mHealth devices in an ethical manner to promote safer use—a literature review. *Studies in Health Technology and Informatics*, *235*, 363−367.

Singh, V., Poonia, R. C., Kumar, S., Dass, P., Agarwal, P., Bhatnagar, V., & Raja, L. (2020). Prediction of COVID-19 coronavirus pandemic based on time series data using support vector machine. *Journal of Discrete Mathematical Sciences & Cryptography*. Available from https://doi.org/10.1080/09720529.2020.1784525.

Slovensky, D. J., Malvey, D. M., & Neigel, A. R. (2017). A model for mHealth skills training for clinicians: Meeting the future now. *Mhealth, 3*.

Spigel, L. W., Wambugu, S., & Villella, C. (2018). *mHealth Data Security, Privacy, and Confidentiality Guidelines: Companion Checklist*. Chapel Hill: MEASURE Evaluation.

Thomas, K. (2018). Wanted: A WhatsApp alternative for clinicians. *British Medical Journal (Clinical Research ed.), 360*, k622.

Tobías, A., Carnerero, C., Reche, C., Massagué, J., Via, M., Minguillón, M. C., ... Querol, X. (2020). Changes in air quality during the lockdown in Barcelona (Spain) one month into the SARS-CoV-2 epidemic. *The Science of the Total Environment, 726*, 138540. Available from https://doi.org/10.1016/J.SCITOTENV.2020.138540.

Vedanthan, R., Blank, E., Tuikong, N., Kamano, J., Misoi, L., Tulienge, D., ... Were, M. C. (2015). Usability and feasibility of a tablet-based Decision-Support and Integrated Record-keeping (DESIRE) tool in the nurse management of hypertension in rural western Kenya. *International Journal of Medical Informatics, 84*(3), 207–219. Available from https://doi.org/10.1016/j.ijmedinf.2014.12.005.

Worldometer (2020). *Coronavirus*, https://www.worldometers.info/coronavirus/country/turkey/.

Yang, Y. T., & Silverman, R. D. (2014). Mobile health applications: The patchwork of legal and liability issues suggests strategies to improve over sight. *Health Affairs, 33*(2), 222–227. Available from https://doi.org/10.1377/hlthaff.2013.0958.

Zeman, J. E., Moon, P. S., McMahon, M. J., & Holley, A. B. (2018). Developing a mobile health application to assist with clinic flow, documentation, billing, and research in a specialty clinic. *Chest, 154*(2), 440–447. Available from https://doi.org/10.1016/j.chest.2018.04.009.

Chapter 4

Combating COVID-19 using object detection techniques for next-generation autonomous systems

Hrishikesh Shenai, Jay Gala, Kaustubh Kekre, Pranjal Chitale and Ruhina Karani
D. J. Sanghvi College of Engineering, Mumbai, India

4.1 Introduction

Object detection is a rapidly growing technology in the field of computer vision, which deals with the detection of objects belonging to a particular class like vehicles, humans, etc. in digital media (videos and images). It is considered as a combination of object localization and image classification techniques as it deals with identifying as well as locating the position of objects belonging to a certain predefined class in the image. The result of object detection algorithms can be displayed either by bounding boxes or by highlighting the pixels belonging to the object found. Approaches based on either deep learning or machine learning are used for object detection. In machine learning-based approaches, each object class is described by some unique features that help in the classification process—for example, spherical balls would appear as a circle in a 2D image. Such predefined features are extracted using feature extraction techniques like the Viola − Jones framework based on Haar features, after which classification is carried out using classification techniques such as support vector machines (SVM) to identify the class. On the other hand, deep learning techniques can perform end-to-end object detection without the need to specify any features, thus saving time. These deep learning methods are generally based on convolutional neural networks (CNN).

The following sections of the chapter explore the architecture and working of some of the most commonly used deep learning-based object detection algorithms: the region-based convolutional neural network (R-CNN) family

Cyber-Physical Systems. DOI: https://doi.org/10.1016/B978-0-12-824557-6.00007-8

and the You Only Look Once (YOLO) family. In the current pandemic situation, these techniques can be of real help in various applications, some of which are described in this chapter. The chapter also proposes a pothole detection system as one of the applications built on YOLOv3, along with results tested on different types of potholes.

4.2 Need for object detection

The sense of vision is one of the most important human senses. Humans can understand the visual scene given in an image or video and gain a lot of information from it. This happens because the human brain detects several entities observed by the eyes and understands the context by analyzing the correlation between these entities. Thus for visual understanding, there is a need to detect the different entities present in visual content. Today, in the age of automation, there is a need for Artificial Intelligence (AI) based systems to perform a multitude of tasks related to vision to reduce the dependency on human workers or, in some cases, to aid them. The AI-based systems need to work similarly to the human brain to detect objects to analyze them, interpret the scene, and perform any further tasks based on the interpretation. Thus object detection is an essential phase of AI-based vision systems. The world is plagued by the COVID-19 pandemic and such AI-based systems can play a major role in helping to mitigate the effects of this pandemic by reducing the dependency on human workers and thus saving lives. Object detection can find its use in various applications that serve to aid in the COVID-19 crisis. Few of the applications are as follows:

- Face mask detector and social distancing detector: To ensure that the citizens are following the safety guidelines in public places.
- COVID-19 detector based on X-rays: To detect the abnormalities in chest X-rays even before the lab reports confirm the clinical symptoms.
- Module for autonomous systems: To help in various sectors to reduce the dependency on human workers, for example, in the transportation sector, where such a system can play a crucial role in the monitoring of road conditions through the use of street cameras. The pothole detection system proposed in the chapter is an example of such a system.

4.3 Object detection techniques

Object detection deals with detecting all the instances of objects present in an image, where these objects belong to particular classes. The object detection models that are based on deep learning contain two major components: an encoder and a decoder. The input image is fed to the encoder to learn and extract the features that are significant for locating and labeling the objects. The encoder output is fed to the decoder to predict the bounding box position and the corresponding label for the object. The metric used to grade the object location predicted by the algorithm is called intersection over union (IOU). Considering the prediction of the model and the ground truth bounding box, compute the area of intersection of the ground truth bounding box and the predicted bounding box, then divide it by the union of the two. The value of IOU is in the range of $0 - 1$. IOU 0 indicates no intersection between the predicted bounding box and ground truth bounding box, while IOU 1 indicates the predicted bounding box is perfectly overlapping with the ground truth bounding box. In this chapter, some of the most commonly used object detection algorithms like the R-CNN family and the YOLO family are described. R-CNN utilizes a selective search algorithm to propose the Region of Interest (RoI) in an image followed by a CNN to detect the presence of the object of interest in those particular regions, thus following a two-step process. YOLO is a single step process where a single neural network is applied to the entire image and then the image is divided into smaller regions and bounding boxes and probability for each region is predicted.

4.3.1   R-CNN family

R-CNN stands for a region-based convolutional neural network. The R-CNN family of object detectors follow two-stage object detector architecture. In a two-stage detector, an RoI pooling layer separates the stages. Fig. 4.1 shows the architecture of a two-stage detector. The candidate object bounding boxes are proposed by a Region Proposal Network (RPN). This acts as the first stage of the detector. While in the second stage, the process of feature extraction is performed by the RoI pool (RoI pooling) operation by

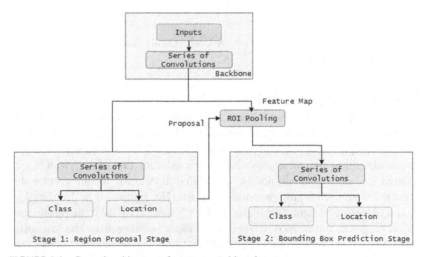

FIGURE 4.1 General architecture of a two-stage object detector.

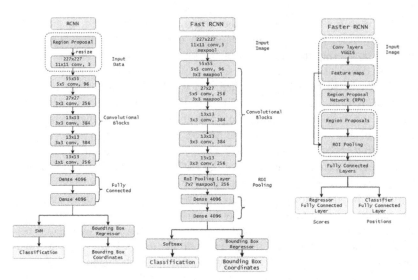

FIGURE 4.2 The network architecture of object detectors in the R-CNN family [R-CNN (left), Fast R-CNN (center), and faster R-CNN (right)].

considering every candidate box for performing further tasks of classification and bounding-box regression. The advantage of using two-stage detectors is high localization and object recognition accuracy.

The main aim of this family is to improve model performance. Fig. 4.2 summarizes the network architecture of object detectors in the R-CNN family (Girshick, Donahue, Darrell, & Malik, 2014; Girshick, 2015; Ren, He, Girshick, & Sun, 2017).

4.3.1.1 R-CNN

R-CNN is the first model in the R-CNN family and was proposed by Girshick et al. (2014).

4.3.1.1.1 Network architecture

A region proposal algorithm is used to identify a feasible number of region proposals in the input image and then feature maps are extracted using CNN from each region independently for classification. R-CNN uses the selective search algorithm to extract just 2000 regions from the image, which are referred to as region proposals. A 4096-dimensional feature vector is extracted from each region proposal using AlexNet (proposed by Krizhevsky et al.). AlexNet requires the input to be of size 227×227; therefore the region proposals are resized irrespective of aspect ratio or size. The last softmax layer present in AlexNet is removed and fully connected (FC) layers are added. The input image is fed to a series of five convolution layers and then passed through two FC layers, which are 4096 dimensional to get a 4096-dimensional feature map. This network architecture is shown in Fig. 4.2. The class of the object, which is represented by the feature vector, is detected by using an SVM per object class (one-vs-one strategy). Each SVM outputs a score for that particular class. This score is an indicator of the likelihood that the region proposal belongs to that particular class. The region proposal is assigned a class label corresponding to the SVM with the highest score (Krizhevsky, Sutskever, & Hinton, 2017).

4.3.1.1.2 Advantages
1. The region proposal algorithm provides a computationally less expensive method of identifying a feasible number of region proposals.

4.3.1.1.3 Disadvantages
1. The SVM is trained on the feature vectors that are generated by AlexNet. That means that training the SVM classifier is not possible before CNN is fully trained. This implies that the training process is not parallelizable.
2. R-CNNs require longer training.
3. For real-time systems, R-CNN is not feasible (it takes 47 seconds for detecting objects in each test image).
4. The selective search algorithm is a fixed algorithm (no learning procedure happens at each stage). This may result in bad candidate region proposals.

4.3.1.2 Fast R-CNN
Fast R-CNN was proposed by Ross Girshick (Girshick, 2015) in 2015 as an improved version of R-CNN. R-CNN was a major development in object detection; however, it had some shortcomings. In R-CNN, for each region proposal, a feature map was calculated, making it slow. Since each feature map was saved, memory requirements were high. The process becomes complicated as a result of separate training of Bounding Box Regressor, CNN, and SVM. Fast R-CNN takes the entire image as an input, from which features are extracted and passed through an RoI pooling layer. This layer outputs a set of fixed-sized feature maps that form the inputs for further classification and localization.

4.3.1.2.1 Network architecture
The Fast R-CNN takes the entire image as an input along with some object proposals. The image is first processed by a network using multiple convolutional layers coupled with max pooling layers to extract a convolutional feature map. The extracted feature map is then passed to the RoI pooling layer. The network architecture is shown in Fig. 4.2.

4.3.1.2.2 The RoI pooling layer

This layer helps to achieve a significant speedup in training and testing while maintaining high accuracy. The layer takes two inputs:

1. Fixed sized feature map from the previous CNN layers.
2. An $N \times 5$ matrix, a list of N RoIs. The first column is the index of the image, while the remaining four columns define the positions of the corners of the bounding boxes.

The layer first divides the region proposal window (of size $h \times w$) into several equal-sized ($H \times W$) subwindows. Max pooling is applied in each of the subwindows, across each channel independently, to generate the corresponding output value. The output dimensions depend only on the number of sections into which the proposal is divided. The benefit of this layer is that even if multiple objects are present in a proposal, the same feature map can be used for each of them, thus reducing the overall processing speed. Each feature vector is then given as an input to a set of FC layers. The output of these layers is forked into the two separate layers. The first layer is a softmax classification layer, wherein the object classes are identified. The second layer is a Bounding Box Regressor layer that generates a tuple of four real-valued numbers, which define the positions of the corners of the bounding box for each of the "K" object classes (Girshick, 2015; Grel, 2017).

4.3.1.2.3 Advantages

1. Fast R-CNN has an improved mean average precision (mAP) over R-CNN.
2. It uses an RoI pooling layer to get a fixed-sized feature map, reducing computation time.
3. It uses truncated Singular Value Decomposition (SVD) to reduce the time required for the computation of FC layers.

4.3.1.2.4 Disadvantages

1. Searching for region proposals takes a lot of time and in the overall running of the algorithm; this slows down the entire architecture.

4.3.1.3 Faster R-CNN

Selective search is utilized by the algorithms discussed earlier (R-CNN and Fast R-CNN) to determine the region proposals. It is heavily time-consuming due to its taxing Central Processing Unit (CPU) implementation, which takes around 2 seconds per image, thereby affecting the performance of the overall network for object detection. Ren et al. (2017), in 2015, proposed a computationally efficient object detection algorithm that unifies the RPNs with the Fast R-CNN. This algorithm reduces the computation time, for determining the region proposals, from 2 seconds to 10 milliseconds improving the performance of the overall network for object detection. In short, the Faster R-CNN algorithm comprises two modules: region proposal by the RPN followed by the Fast R-CNN detector.

4.3.1.3.1 Network architecture

RPN: The RPN takes an input image and then outputs the region proposals along with their objectness score. The RPN and Fast R-CNN share a common set of convolutional layers. The network has to determine whether an object is present in the input image by generating the region proposals by sliding a small network over the convolutional feature map obtained from the last shared convolutional layer. The sliding window is mapped to a lower-dimensional feature subspace (512-d for VGGNet and 256-d for ZFNet). The feature space is fed into the two FC layers—a regression layer (reg) and a classification layer (cls). As the network slides through the convolutional feature map, it checks whether "k" corresponding anchors contain an object and refines the anchor coordinates to give the bounding boxes. Fig. 4.3 shows the RPN (Ananth, 2019; Simonyan & Zisserman, 2015; Szegedy et al., 2015; Xia, Chen, Wang, Zhang, & Xie, 2018; Zeiler & Fergus, 2014).

For k anchors boxes, the regression layer outputs 4k coordinates of the bounding boxes and the classification layer outputs 2k scores that estimate the probability of the presence of an object for each region proposal. An

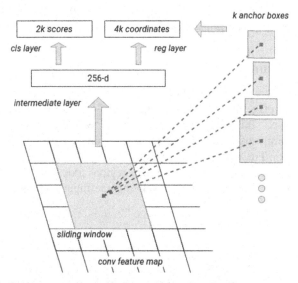

FIGURE 4.3   Region proposal network. *Reprinted from Xia, D., Chen, P., Wang, B., Zhang, J., & Xie, C. (2018). Insect detection and classification based on an improved convolutional neural network, Agricultural Sensing and Image Analysis, 18, 2018, licensed under CC by 4.0.*

anchor is assigned a *positive* label if it satisfies either of the two conditions: (1) the anchors with the highest IOU, that is, a measure of overlap with the ground truth box or (2) an anchor that has an IOU greater than 0.7. An anchor is assigned a *negative* label if its IOU value is lower than 0.3. The anchors that do not satisfy either of the conditions are discarded.

Detector: The convolutional layers are shared by both RPN and Fast R-CNN. The region proposals from the RPN are passed through the RoI pooling layer to obtain feature vectors. These feature vectors are then fed into the sibling classification and regression branches, which are different from those in the RPN. The classification layer estimates the probability of the proposal belonging to a particular class and the regression layer outputs the coordinates of the predicted bounding boxes whose size is specific to each class. The network architecture is shown in Fig. 4.2.

4.3.1.3.2 Advantages

1. It eliminates the CPU-based selective search, which makes the overall network faster.
2. It uses a shared convolutional layer for the RPN and Fast R-CNN instead of separate convolutional networks.

4.3.1.3.3 Disadvantages

1. To extract all the objects from an image, more than one pass is required through a single image.
2. It consists of different modules working one after the other, so the overall performance is proportional to the performance of the previous modules.

4.3.2 YOLO family

YOLO is the abbreviation for "You Only Look Once," which suggests that given an input image, it is possible to detect and localize the objects present in a single glance. This technique treats object detection as a regression task. The YOLO family of object detectors follows one-stage detector architecture, as shown in Fig. 4.4.

One-stage detectors do not have the region proposal step. Thus the bounding boxes are predicted by considering the input images directly and

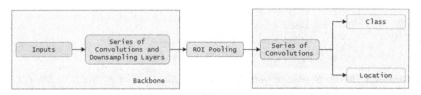

FIGURE 4.4 General architecture of an one-stage object detector.

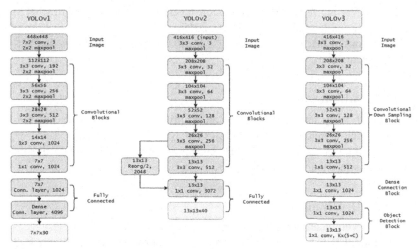

FIGURE 4.5 The network architecture of object detectors in the YOLO family [YOLOv1 (left), YOLOv2 (center) and YOLOv3 (right)].

so a separate region proposal step is not needed. One-stage detectors are efficient concerning time because of their high inference speed and are thus highly suitable for real-time devices. Fig. 4.5 summarizes the network architecture of object detectors in the YOLO family (Redmon & Farhadi, 2018; Redmon, Divvala, Girshick, & Farhadi, 2016; Seong, Song, Yoon, Kim, & Choi, 2019).

4.3.2.1 YOLOv1

YOLO, as the name suggests, aims to look at the image once to detect and localize the objects present in the image. It was proposed by Joseph Redmon et al. (Redmon et al., 2016) in 2015. It has a simple architecture that consists of a single convolutional network that predicts objects across various classes simultaneously by using feature maps obtained from the entire input image. This design helps to provide end-to-end training with real-time speeds while achieving high accuracies. YOLO splits the input image in an $S \times S$ grid. Each cell of this grid is responsible for detecting an object if the object center is present in it. Each cell predicts "B" number of bounding boxes and also generates an objectness score for it. This score reflects the accuracy of the detection of the cell. The objectness score will be zero if no object is in the cell; otherwise, it is the IOU between the ground truth box and the predicted box.

4.3.2.1.1 Network architecture

- YOLO is essentially based on GoogLeNet. It has 24 convolutional layers, which are followed by a couple of FC layers. The inception modules of

GoogLeNet are replaced by reduction layers (1×1) along with (3×3) convolutional layers (Szegedy et al., 2015).

- The output of the network is a ($7 \times 7 \times 30$) tuple. Fig. 4.5 shows this network architecture.

4.3.2.1.2 Advantages

1. It is extremely fast. The images can be processed at 45 frames per second, which is much faster than Faster R-CNN.
2. Background errors made by Fast R-CNN is about 13.6%, while that of YOLO is less than 4.75%.

4.3.2.1.3 Disadvantages

1. It is not very accurate despite being very fast.
2. The model fails to detect small objects in groups accurately (like a bird flock).
3. The model has difficulty in generalizing objects with unusual configurations or aspect ratios.

4.3.2.2 YOLOv2

This version of YOLO focuses mainly on improving localization and recall (a ratio of true object detections to the total number of objects in the data) while also maintaining classification accuracy. It was proposed by Redmon and Farhadi (2017) in 2016.

4.3.2.2.1 Improvements made over YOLOv1

- Batch normalization: It is used on all convolutional layers. Results in a 2% improvement in mAP.
- High-resolution classifier: YOLOv2 uses additional 416×416 images for fine-tuning the classification network. This is carried out for 10 epochs on ImageNet. Results in 4% improvement in mAP.
- Convolutions with anchor boxes: All FC layers are removed and anchor boxes are used to predict bounding boxes. With anchor boxes, recall is 88% and a mAP of 69.2%.
- Direct location prediction: Logistic activation σ is used for location binding. This makes the value fall from 0 to 1. This results in 5% improvement in mAP.
- Fine-grained features: Large objects can be detected using a 13×13 feature map. While to detect smaller objects properly, the maps from the earlier layer are mapped into a $13 \times 13 \times 2048$ feature map and then concatenated with the original 13×13 feature maps. This results in 1% improvement in mAP.

4.3.2.2.2 Network architecture

- Darknet-19 acts as a backbone for YOLOv2. It has 19 convolutional and five max pooling layers.
- Here the output shape is $(13, 13, (k(1 + 4 + 20)))$. As the number of classes is 20 while the number of anchor boxes is denoted by k. Fig. 4.5 shows this network architecture.

4.3.2.2.3 Advantages

1. It has considerable improvement in the IOU score.
2. The mAP was also improved.

4.3.2.2.4 Disadvantages

1. It is not suitable for a real-time production system due to its slowness.
2. It cannot detect smaller objects and objects with unusual aspect ratios.

4.3.2.3 YOLOv3

YOLOv2 had certain drawbacks such as (1) not good at detecting small objects, (2) assuming that the object classes are mutually exclusive. Redmon and Farhadi (2018) in 2018 proposed a new architecture for the YOLO known as YOLOv3, which tries to solve the problems faced by YOLOv2 architecture by collectively accumulating good ideas from other architectures. YOLOv3 boosts the accuracy of the overall network.

4.3.2.3.1 Improvements made over YOLOv2

- Bounding box prediction: For every bounding box, logistic regression is used to determine the "objectness score."
- In case of overlap between the preceding bounding box and a ground truth object, we get an objectness score of "1." Only one bounding box gets assigned for each of the ground truth objects.
- Class prediction: YOLOv3 eliminates the softmax classification and uses the multilabel classification for predicting the classes for the bounding boxes. During training, a binary cross-entropy loss is used for the class predictions.
- Prediction across scales: YOLOv3 predicts bounding boxes by extracting the features at three different scales, much similar to the Feature Pyramid Network. Each bounding box prediction comprises four bounding offsets, 1 objectness score and 80 class scores. K-means clustering is used for determining the bounding boxes (Lin et al., 2017).

4.3.2.3.2 Network architecture

YOLOv3 uses a hybrid Darknet-53, which consists of 53 layers of Darknet and an additional 53 layers leading to a total of 106 layers. Fig. 4.5 shows this network architecture.

4.3.2.3.3 Advantages

1. Average precision is improved for the small objects and is better than Faster R-CNN.
2. mAP was increased significantly and localization errors were reduced.
3. Predictions at different scales improved.

4.3.2.3.4 Disadvantages

1. Average precision for the large and medium objects can be improved.
2. mAP score between 0.5 and 0.95 IOU can still be increased.

4.4 Applications of objection detection during COVID-19 crisis

Object detection is useful in various applications, which will be crucial in combating the COVID-19 crisis. Few of the applications are as follows.

4.4.1 Module for autonomous systems (pothole detection)

The pandemic has affected various sectors that are dependent on physical labor. With lockdowns being imposed for curbing the spread of this deadly disease, the sectors that depend on physical human labor have been affected the most. One such area is the transportation sector, where the road conditions have to be monitored and maintained continuously. An AI-based autonomous system would play a crucial role in the monitoring of road conditions through the use of surveillance cameras where the road anomalies can be detected using the object detection module. This module would detect potholes using encoded representation of images. As most of the cities have now become "smart cities," most of the roads are covered by surveillance cameras spanning across the city. The YOLO family is used here since pothole detection is a real-time system and the R-CNN family, though more accurate, is slower as compared to the YOLO family.

4.4.1.1 Architecture

Dataset used: Dataset consists of 1500 images taken with a mobile phone mounted on the dashboard of a car. Images are captured with different lighting conditions and climatic conditions and are annotated according to YOLO standards using an annotation tool. The train test split is 80−20.

Network architecture: YOLOv3 from the YOLO family is used as the network architecture. The architecture is based on the pretrained weights of a hybrid Darknet-53. It is tuned to obtain the required mAP value. Parameters like batch size and subdivisions can be tweaked to match the training system's capabilities. The filter size is 18×18 (Redmon & Farhadi, 2018).

4.4.1.2 Results

The proposed system gives an average IoU value of 0.72 while the YOLOv2 model has a value of 0.61 (both trained on the above-mentioned dataset). Fig. 4.6A and B show pothole detection achieved with the help of YOLOv3 through experiments conducted.

4.4.2 Social distancing detector

COVID-19 spreads among people who stay nearby (about 6 feet) for long periods. The virus spreads when an infected person coughs, sneezes, or even talks and droplets containing the virus are launched into the air. These droplets can travel a distance of about 6 feet and reach the mouth or nose of other people and get inhaled into their lungs. These infected people may not show symptoms; however, they play a role in the spreading of the virus. Social distancing helps to restrict the opportunities of coming in contact with infected people outside the home. It is important since this would reduce the spread of the virus. Social distancing has to be practiced strictly to stop the spread of this virus. It has been found through studies that the spread of this virus is not dependent on the age of the person and thus it is of paramount importance that social distancing is practiced by every single person (Bhatnagar et al., 2020). Several police officials have to risk their lives every day by being out in the field to make sure that social distancing is being maintained. Social distancing detector can help such officers by allowing them to monitor various public places remotely. Object detection would be an important module to detect people and analyze the distance between them. If the distance between the detected people is below a particular threshold then the respective

FIGURE 4.6   A. Waterlogged pothole. B. Normal pothole.

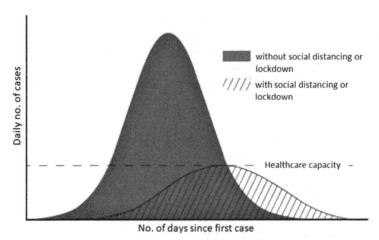

FIGURE 4.7   Effect of social distancing on the reduction of the number of COVID-19 cases. *Reprinted from Punn, N.S., Sonbhadra, S.K., & Agarwal, S. (2020) Monitoring COVID-19 social distancing with person detection and tracking via fine-tuned YOLO v3 and Deepsort techniques, 2020, e-print arXiv:2005.01385v2, arXiv., licensed under CC by 4.0.*

authorities would be alerted to take the necessary measures. Fig. 4.7 clearly describes how social distancing has helped in reducing the number of positive cases (Punn, Sonbhadra, & Agarwal, 2020).

Dataset used: Open image dataset. Images are resized such that the shorter edge contains "P" pixels, which range from 600 for a low resolution to 1024 for high resolution. Along with this, surveillance footage from Oxford Town Center is utilized, which is also further used for demonstrating the overall working of the module (Google, 2020; Wojke, Bewley, & Paulus, 2017).

Approach: Faster R-CNN model or YOLOv3 can be used for object detection with Deepsort for tracking objects surrounded by bounding boxes. The bounding boxes are then analyzed to identify people not following social distancing. Along with this, every bounding box is color-coded to keep track of people of the same group by having the same colored bounding boxes for them. The output also consists of a streamline plot showing the statistical data like the number of groups and ratio of the number of people to the number of groups, which can be helpful to determine the level of risk at a particular place. The workflow of the system is as follows:

1. Each detected person is associated with a tuple (x, y, d) where (x, y) is the centroid of the coordinates of the bounding boxes, d is the approximate depth of that person as viewed from the camera.
2. The pairwise L2 norm is then calculated. This is used to determine the closeness of an individual with each of the neighbors. The closeness threshold is updated constantly.
3. The color of the bounding box of a person is changed when they satisfy the closeness criteria of some other person, thereby indicating the

FIGURE 4.8 Working demo of the proposed model on video surveillance of Oxford Town Center. *Reprinted from Punn, N.S., Sonbhadra, S.K., & Agarwal, S. Monitoring COVID-19 social distancing with person detection and tracking via fine-tuned YOLO v3 and Deepsort techniques, 2020, e-print arXiv:2005.01385v2, arXiv, licensed under CC by 4.0.*

formation of groups. This indicates a violation of social distancing norms (Punn et al., 2020).

4.4.2.1   Results

The proposed system using faster R-CNN gives an mAP value of 0.969, while YOLOv3 has a value of 0.846 (Punn et al., 2020).

Fig. 4.8 shows the working of the model where the first image has a violation index of 3 (presence of larger groups of people) and the second Image has a violation index of 2 (fewer people in immediate proximity) (Punn et al., 2020).

4.4.3   COVID-19 detector based on X-rays

Most of the severe health conditions are observed as a result of the late detection of the infection. Because of the tremendous pressure on the healthcare department, there is a delay in the receipt of the actual laboratory diagnosis reports leading to the deterioration of the health condition of the patient. The test samples collected are being sent to the specialized laboratories in urban areas and this leads to a considerable delay, especially for rural areas. Had this delay been reduced, some precautionary medication could have been administered early, which might have helped in ameliorating the patient's health condition. Moreover, early detection would have helped ensure that the patient is quarantined, thus restricting the further spread of the virus. X-ray machines are already present in most of the healthcare systems available in rural areas and by using these X-ray systems, the transportation time can also be saved as well as the pressure on the laboratories would be reduced to some extent. An AI-based COVID-19 detector can be developed, which could detect the abnormalities even before the lab reports confirm the clinical symptoms to detect the presence of the virus based on X-ray imagery. An object detection module is needed in such AI systems that will detect abnormalities from X-ray imagery that could be indicators of the virus. This will not only help the doctors administer immediate medical aid at an initial stage but also help to curtail the spread of the virus.

4.4.3.1 Architecture

Dataset used: Dataset prepared by Cohen, Morrison, and Dao (2020) by collecting X-ray images from various sources (open-access). This dataset contains chest X-ray images of 82 male patients and 43 female patients who were diagnosed as positive cases. Few samples from the ChestX-ray8 dataset that has been provided by Wang et al. (2017), which consists of chest X-ray images of normal patients and pneumonia patients, have been used.

DarkCovidNet: Darknet-19 model is used as a starting point. Each DarkNet layer consists of one convolutional layer, which is followed by BatchNorm and Leaky Rectified Linear Unit (ReLU) activation operation. Max pooling is used in all the pooling operations. The learning rate of $3e^{-3}$ is considered. A cross-entropy loss function is used along with Adam optimizer for updating the model parameters.

4.4.3.1.1 Results

The proposed system has an accuracy of 98.08%. The model highlights imperfections in the X-ray images using heatmaps. Fig. 4.9 shows the heatmaps of corresponding chest X-ray images (Ozturk et al., 2020).

4.4.4 Face mask detector

Face masks act as the first line of defense against this deadly virus. Wearing a face mask prevents the spread of infection while also protecting the users. Face masks eliminate cross-contamination. Automatic face mask detection systems can help governments to monitor the citizens and ensure that they are taking necessary precautions (wearing face masks when going in public places) to ensure that the virus is not spreading. Object detection can be used

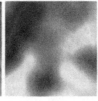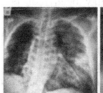

FIGURE 4.9 X-ray images and corresponding heatmaps. *Reprinted from Ozturk, T., Talo, T., Yildirim, E.A., Baloglu, U.B., Yildirim, O., & Acharya, U.R. (2020). Automated detection of COVID-19 cases using deep neural networks with X-ray images,* Computers in Biology and Medicine, 121, *1–11.*

FIGURE 4.10   No face mask detected (left) and face mask detected (right).

for detecting the presence of face masks on faces of the people, thereby proving if they are wearing a face mask or not (Jeremy et al., 2020).

4.4.4.1   Architecture

Dataset used: Face dataset of 1376 images divided between two classes:

- Without_mask: 686 images
- Mask: 690 images

The method used for creating the dataset:

1. Take normal face images (without masks).
2. Use a custom computer vision python script to append face masks on them.

The only precaution to be taken is that the images used to create the "mask" dataset cannot be used in the "Without_mask" dataset. The model becomes biased and would not generalize well if the original images are used as "Without_mask" samples (Rosebrock, 2020).

Network architecture: The proposed system uses YOLOv3 from the YOLO family for face mask detection. It uses DarkNet-53 for feature detection (Rodriguez & Lorenzo, 2020).

4.4.4.1.1   Results

Fig. 4.10 shows the results of face mask detection using YOLOv3 (Rodriguez & Lorenzo, 2020).

4.5   Conclusion

Object detection is a widely growing computer vision technique that is essential for understanding and analyzing visual scenes in a video or a photo.

It involves two processes: identifying and locating objects of different predefined classes. Deep learning-based approaches use CNNs to perform the task of object detection in an end-to-end, unsupervised manner. R-CNN and YOLO families, as discussed in the chapter, are popularly used for object detection. Each family has some trade-offs between speed and accuracy. R-CNN family algorithms have higher accuracy but slower speeds than the YOLO family, whereas YOLO family algorithms are computationally faster but have slightly lower accuracy than the R-CNN family. Object detection can be helpful in mitigating the transmission of the COVID-19 virus by automating monitoring systems, thereby reducing human dependence. Few other applications of object detection include self-driving cars, video surveillance, and anomaly detection.

References

Ananth, S. (2019) *Faster R-CNN for object detection*. Retrieved from https://towardsdatascience. com/faster-r-cnn-for-object-detection-a-technical-summary-474c5b857b46,Medium.

Bhatnagar, V., Poonia, R. C., Nagar, P., Kumar, S., Singh, V., Raja, L., & Dass, P. (2020). Descriptive analysis of COVID-19 patients in the context of India. *Journal of Interdisciplinary Mathematics*, 1−16. Available from https://doi.org/10.1080/09720502.2020.1761635.

Cohen, J.P., Morrison, P., & Dao, L. (2020). COVID-19 image data collection, e-print arXiv:2003.11597, arXiv.

Girshick, R. (2015). Fast R-CNN. *International Conference on Computer Vision (ICCV)*, 1440−1448. Available from https://doi.org/10.1109/ICCV.2015.169,IEEE.

Girshick, R., Donahue, J., Darrell, T., & Malik, J. (2014). Rich feature hierarchies for accurate object detection and semantic segmentation. *Conference on Computer Vision and Pattern Recognition (CVPR)*, 580−587. Available from https://doi.org/10.1109/CVPR.2014.81,IEEE.

Google. (2020). Open image dataset v6, Retrieved from https://storage.googleapis.com/openimages/web/index.html,Google.

Grel, T. (2017). *Region of interest pooling explained*. Retrieved from https://deepsense.ai/region-of-interest-pooling-explained,deepscience.ai.

Jeremy, H., et al. (2020). Face masks against COVID-19: An evidence review. *Preprints*, *30* (20), 1−9. Available from https://doi.org/10.20944/preprints202004.0203.v2.

Krizhevsky, A., Sutskever, I., & Hinton, G. (2017). ImageNet classification with deep convolutional neural networks. *Communications of the ACM*, *60*(6), 84−90. Available from https://doi.org/10.1145/3065386,ACM.

Lin, T. Y., Dollar, P., Girshick, R., He, K., Hariharan, B., & Belongie, S. (2017). Feature pyramid networks for object detection. *Conference on Computer Vision and Pattern Recognition (CVPR)*, 2117−2125. Available from https://doi.org/10.1109/CVPR.2017.106.

Ozturk, T., Talo, M., Yildirim, E. A., Baloglu, U. B., Yildirim, O., & Acharya, U. R. (2020). Automated detection of COVID-19 cases using deep neural networks with X-ray images. *Computers in Biology and Medicine*, *121*, 1−11. Available from https://doi.org/10.1016/j.compbiomed.2020.103792, ScienceDirect.

Punn, N.S., Sonbhadra, S.K., & Agarwal, S. (2020). Monitoring COVID-19 social distancing with person detection and tracking via fine-tuned YOLO v3 and Deepsort techniques, e-print arXiv:2005.01385v2, arXiv.

Redmon, J., & Farhadi, A. (2017). YOLO9000: Better, faster, stronger. *Conference on Computer Vision and Pattern Recognition (CVPR)*, 6517−6525. Available from https://doi.org/ 10.1109/CVPR.2017.690.

Redmon, J., & Farhadi, A. (2018). YOLOv3: An incremental improvement, preprint arXiv:1804.02767, arXiv.

Redmon, J., Divvala, S., Girshick, R., & Farhadi, A. (2016). You Only Look Once: Unified, real-time object detection. *Conference on Computer Vision and Pattern Recognition (CVPR)*, 779−788. Available from https://doi.org/10.1109/CVPR.2016.91, IEEE.

Ren, S., He, K., Girshick, R., & Sun, J. (2017). Faster R-CNN: Towards real-time object detection with region proposal networks. *Transactions on Pattern Analysis and Machine Intelligence, 39* (6), 1137−1149. Available from https://doi.org/10.1109/TPAMI.2016.2577031,IEEE.

Rodriguez, L., & Lorenzo, A. (2020) Face mask detector using deep learning (YOLOv3). Retrieved from https://medium.com/face-mask-detector-using-deep-learning-yolov3/face-mask-detector-using-deep-learning-yolov3-209b57f77e92,Medium.

Rosebrock, A. (2020). COVID-19: Face mask detector with OpenCV, Keras/TensorFlow and deep learning. Retrieved from https://www.pyimagesearch.com/2020/05/04/covid-19-face-mask-detector-with-opencv-keras-tensorflow-and-deep-learning,Pyimagesearch.

Seong, S., Song, J., Yoon, D., Kim, J., & Choi, J. (2019). Determination of vehicle trajectory through optimization of vehicle bounding boxes using a convolutional neural network. *Sensors (Basel)., 19*(01), 1−18. Available from https://doi.org/10.3390/s19194263, Sensors.

Simonyan, K., & Zisserman, A. (2015). Very deep convolutional networks for large-scale image recognition. *International Conference on Learning Representations (ICLR)*, preprint arXiv:1409.1556, arXiv.

Szegedy, C., Liu, W., Jia, Y., Sermanet, P., Reed, S., Anguelov, D., ... Rabinovich, A. (2015). Going deeper with convolutions. *Conference on Computer Vision and Pattern Recognition (CVPR)*, 1−9. Available from https://doi.org/10.1109/CVPR.2015.7298594.

Wang, X., Peng, Y., Lu, L., Lu, Z., Bagheri, M., & Summers, R. M. (2017). ChestX-Ray8: Hospital-scale chest X-ray database and benchmarks on weakly-supervised classification and localization of common thorax diseases. *Conference on Computer Vision and Pattern Recognition (CVPR)*, 3462−3471. Available from https://doi.org/10.1109/CVPR.2017.369.

Wojke, N., Bewley, A., & Paulus, D. (2017). Simple online and realtime tracking with a deep association metric. *International Conference on Image Processing (ICIP)*, 3645−3649. Available from https://doi.org/10.1109/ICIP.2017.8296962.

Xia, D., Chen, P., Wang, B., Zhang, J., & Xie, C. (2018). Insect detection and classification based on an improved convolutional neural network. *Agricultural Sensing and Image Analysis, 18*. Available from https://doi.org/10.3390/s1812416910.3390/s18124169.

Zeiler, M. D., & Fergus, R. (2014). Visualizing and understanding convolutional neural networks. *European Conference on Computer Vision (ECCV)*, 818−833. Available from https://doi.org/10.1007/978-3-319-10590-1_53.

Chapter 5

Non-contact measurement system for COVID-19 vital signs to aid mass screening—An alternate approach

Vijay Jeyakumar, K. Nirmala and Sachin G. Sarate

Department of Biomedical Engineering, Sri Sivasubramaniya Nadar College of Engineering, Chennai, India

5.1 Introduction

The world is seeing an outbreak of a big pandemic. The record of events that lead to the available information about this disease goes like this. A sudden appearance of a large number of cases with lower respiratory tract infections with an unknown disease was detected in Wuhan city. This was reported to the World Health Organization (WHO) country office in China. Wuhan is the largest metropolitan area in China's Hubei province. The beginning of symptomatic individuals of this disease can be traced through published literature to the start of December 2019. The earlier cases were classified as "pneumonia of unknown etiology" because the causative agent was unidentified. The number of cases was 29. The Chinese Centers for Disease Control and Prevention (CDC) and the local CDCs ran a program to investigate this event. They attributed the etiology of this disease to a new coronavirus. The Director-General of WHO announced on February 11, 2020, that the disease caused by this new virus was coronavirus disease 2019 and an acronym COVID-19 was coined. Severe Acute RAR Respiratory Syndrome Coronavirus (SARS-CoV) and Middle East Respiratory Syndrome Coronavirus (MERS-CoV) are the two other CoV epidemics that have occurred in the past 20 years. The former one had a fatality rate of 9.6% with about 8000 cases and 800 deaths and the later one began in Saudi Arabia, which had a fatality rate of 35% with about 2500 cases and 800 deaths.

The virus has spread globally in a very short period as it is very contagious. The WHO declared it as Public Health Emergency of International

Cyber-Physical Systems. DOI: https://doi.org/10.1016/B978-0-12-824557-6.00006-6
75

Concern as it had disseminated to 18 countries. Out of these 18, four countries had reported human to human transmission. This declaration was done on January 30, 2020. On February 26, 2020, the first case of disease, which was not related to China, was recorded in the United States.

The name given to the virus was 2019-nCoV in the initial stage. The International Committee on Taxonomy of Viruses gave a new name to it— the SARS-CoV-2 virus. This was done because the virus is similar to the one that caused the SARS outbreak (SARS-CoVs).

This virus belongs to a large family of viruses that have a single-stranded RNA. They can be isolated from a varied number of animal species. This family of viruses has become the dominant pathogen of looming respiratory disease epidemics. These viruses can shift between species and can cause diseases like mild common cold to more harsh diseases. The dynamics of SARS Cov-2 are not known exactly and it is being speculated that it also has an animal origin like SARS-CoV, which came from Himalayan palm civet and MERS-CoV, which came from camels to humans.

The introduction to COVID-19 is discussed in section 5.1 followed by the global scenarios of COVID-19 in section 5.2 of this chapter. The various measurement tools and testing protocols are briefed in section 5.3. The non-contact approach to physiological measurement, the proposed methodology and its experimental results are discussed in section 5.4.

5.2 COVID-19 global scenarios

On February 28, 2020, the WHO raised the threat to the COVID-19 epidemic to "very high." On March 11, 2020, the disease was declared as pandemic because the number of cases outside China had increased 13 times. Around 114 countries were involved, and the number of cases crossed 118,000.

It has been estimated that strict shutdowns at the beginning of the disease would have saved 3 million lives across 11 European countries. All over the world, governments are working hard to establish measures to stop the effects. To reduce the impact of the threat, health organizations are streamlining the flow of information and issuing directives and guidelines. Scientists all over the world are working hard to know the mechanism of transmission, the clinical picture, new ways to diagnose the disease, its prevention, and are building new treatment protocols. It is still uncertain as to when this pandemic will reach its peak.

5.2.1 Infections, recovery and mortality rate

Even until now, the treatment methods are only of supportive form. The best weapon is prevention with a motive to reduce human to human transmission in the community. It was seen that the cause of the reduction in cases in China was the implementation of aggressive measures to isolate individuals.

The disease had spread to Europe from China. The disease spread in Italy, started in the northern region and then throughout the country. It was a testing time for the health system in Italy. Humongous efforts were taken by political and health authorities.

The disease then spread to the United States very quickly. Afterward, the COVID-19 quickly crossed the ocean and as of August 15, 2020, about 21,487,828 cases (with 766,027 deaths) have been recorded in the United States, whereas Brazil with more than 3,282,101 cases and about 106,608 deaths is the most affected state in South America and the second in the world after the United States. India has become the third most affected country with 2,587,461 cases and 50,080 death cases. Across the globe, the total number of COVID-19 affected (21,487,282), recovered (14,243,436), and death (766,027) cases are increasing on an hourly basis.

However, the lethality rate of this disease has been seen to be significantly less than the other two Co-V diseases. But the spread of the SARS-CoV-2 virus across the continents is much faster than any other viruses before. Due to previous underlying conditions like diabetes and hypertension, the estimate says that about one in five individuals could be at risk worldwide.

5.2.2 Economy and environmental impacts

Migrant workers are involved in jobs that are avoided by urban natives. Due to poverty, migrant workers are vulnerable to measures like a sudden lockdown. During the lockdown in India, it was observed that because of being excluded in urban society, migrant workers did not have access to social security programs and healthcare. The sudden lockdown imposed upon the country made many migrant workers to be stranded in different cities and those who were traveling became grounded at stations or state and district borders.

A large number of these workers had to walk to reach their home villages due to lack of public transport. Even after reaching home, they were seen as a risk by the locals reported by R.B. Bhagat, Reshmi, Sahoo, Roy, and Govil (2020). A study made on the effect of COVID-19 on freight market dynamics in Germany used correlation and regression analysis and found out that the growth of transport volume was influenced by COVID-19, depending on the number of new cases and the number of deaths per day as observed by Dominic Loske (2020).

A review by Nicolaa et al. (2020) studied the impact of COVID-19 on various sectors like agriculture, petroleum and oil, manufacturing and industry, education, finance, healthcare and pharmaceutical, hospitality, real estate and housing, sports, information technology, media, research, and development as well as food sector. They also speculated on the response given by Europe, the United Kingdom, the United States, China, and Japan. They observed that unemployment was caused because of measures taken by various countries like social distancing, self-isolation, and restriction of travel. The demand for commodities and manufactured products has been reduced a lot. However, there is a significant increase in the need for medical supplies. Because of the panic buying and stockpiling of food products, the food sector is also facing an increase in demand.

One study demonstrated the use of a linear input − output model to make an estimate of unemployment caused in India due to COVID-19. This study observed that the Indian economy will lose a significant amount of its GDP, around 10%−31%. It was also observed in this study that the daily supply of power from coal-fired power plants reduced by 26% and the CO_2 emissions were also reduced correspondingly to about 15−65 $MtCO_2$. The emission requirements imposed on coal-fired power plants will further harm the sector, suggests the analysis.

5.3 Measurement and testing protocols of COVID-19

The advent of a new virus means there is only a poor knowledge of transmission mechanisms, immunity, seriousness, clinical manifestations, and comorbidity for infection. To address the new virus, the WHO has launched a global initiative to allow any country in any resource setting to collect rapidly robust data on key epidemiological parameters to identify, respond to, and monitor the COVID-19 pandemic. These protocols are designed to collect and share data quickly and systematically in a manner that enables the collection, tabulation, and analysis across various global settings. WHO also developed several standardized investigation protocols in collaboration with technical partners, called the World Health Organization (WHO), 2020 WHO Unity Studies that includes a mixture of serological and molecular tests. The widespread adoption of Unity protocols in more than 90 countries illustrates their effectiveness in improving decision-making across countries to implement or lift effective public health and social strategy to prevent and manage COVID-19 infection.

The testing protocols framed by the WHO are

1. Population-based age-stratified seroepidemiological investigation for the general population.
2. The first few COVID-19 X cases and contacts transmission investigation protocol (FFX).
3. Household transmission of COVID-19.
4. Assessment of COVID-19 risk factors among health workers.
5. Protocol for a case study control.
6. Schools and other educational institutions transmission investigation protocol for COVID-19.
7. Surface sampling of COVID-19 virus.

5.3.1   Measurement methods

Diagnosis is based, as with other respiratory viruses, on two major components: clinical indications such as fever, nausea, dry cough, shortness of breath, and gastrointestinal symptoms. The procedures used for paraclinical diagnosis range from polymerase chain reaction (PCR) to computed tomography (C.T.) as reported by Longb et al. (2020) and Kumari et al. (2020). A fast and reliable diagnosis is in high demand in such a pandemic situation. There are two key populations at higher risk of acquiring a severe illness, including the elderly and those with existing chronic health problems such as diabetes mellitus, obesity, cardiorespiratory disorders, chronic liver diseases, and renal failure. A study was carried out to find the relationship between gender and transmission type in the patients by R. Kumari et al. (2020) and Bhatnagar et al. (2020). Cancer patients and those with immunosuppressive medications as well as pregnant women also have a high chance of transmitting severe illnesses when infected. Current research suggests that nausea, fatigue, cough, and productive cough were the most frequently reported symptoms in COVID-19 patients as seen by V. Singh et al. (2020) and Jieyun Zhu et al. (2020). In these patients, important physiological measurements include respiratory rate, oxygen saturation, heart rate (HR), blood pressure, temperature, level of consciousness using the Glasgow Coma Scale, pain score, and urine production. In the next section, we will discuss the diagnostic tools deployed to help manage the outbreak. Two major categories of diagnostic tools include pathophysiological tools and respiratory assessment tools.

5.3.1.1   Pathophysiological tools

In this method, suspected cases should be tested for confirming the presence of the virus with nucleic acid amplification tests (NAAT), such as reverse transcription polymerase chain reaction (RT-PCR). Infection with laboratory-confirmed SARS-CoV-2 requires real-time detection of viral nucleic acid in samples of the respiratory tract using RT-PCR assay. In the case of ambulatory patients, the samples are collected from the nasopharyngeal and oropharyngeal tract using a swab. Whereas the patient with more severe respiratory conditions sputum from the lower respiratory tract and the lavage from the endotracheal aspirate or bronchoalveolar are collected and tested. After the collection of the sample for virus detection, it should be ensured that it reaches the laboratory as soon as possible. The collected samples can be stored and transported at $2-8°C$.

5.3.1.1.1   Nucleic acid amplification tests

The standard detection of COVID-19 cases is focused upon NAAT that detects unique sequences of RNA viruses, as the real-time RT-PCR confirms the nucleic acid sequencing. A positive NAAT result for at least two distinct targets on the COVID-19 virus genome, of which at least one target is ideally unique to COVID-19 virus using a reliable assay and COVID-19 virus accordingly detected by decoding part or whole of the virus genome as long as the target molecule is larger or different than the amplicon tested in the NAAT assay used. The negative result in the infected person may be due to the poor quality of the specimen, less quantity of the sample, improper handling of the sample or late collection of sample.

5.3.1.1.2 Serological testing

The evaluation of the infected patient and the magnitude of the outbreak can be investigated by the serological survey. When the NAAT assay test is negative, the validated serology test helps to diagnose the paired serum sample to confirm the presence of the virus. Serum samples were tested using the enzyme-linked immune sorbent assay. Cross-reactivity to other coronaviruses can be very difficult, but commercial and non-profit serological studies are currently being developed and evaluated.

5.3.1.2 Physiological assessment tools

The essential physiological observations to decide whether a person is infected by COVID-19, include temperature, respiratory rate, oxygen saturation, pulse rate, blood pressure, and chest imaging as reported by Carter, Aedy, and Notter (2020). Temperature measurement may be a part of the evaluation to decide whether a person has a high temperature potentially caused by an infection with COVID-19. A pulse oximeter is used to measure the blood oxygen levels in patients with preexisting respiratory conditions. In this pandemic situation, the pulse oximeter helps to measure the gradual depletion of the oxygen level in the asymptomatic individual. Because of the asymptomatic nature of most COVID-19 infections, many patients with coronavirus may suffer from significantly lower blood saturation levels without even recognizing it. The oxygen saturation of a person normally ranges from 95% to 100% (SpO_2) as evaluated by the pulse oximeter and requires immediate assistance if it begins to fall below 92.

The chest X-rays of patients infected with the novel coronavirus show characteristic pneumonia-like patterns that can help in the diagnosis. A chest X-ray can be used as an efficient, simple, fast, and inexpensive way to quickly diagnose the COVID-19 suspected individual before the confirmation from the pathophysiological test, namely the PCR test. While chest C.T. scans offer more reliable features for COVID diagnosis, they are not readily available in resource-limited environments.

5.3.2 COVID-19 innovations

COVID-19 is a fairly new disease; therefore, it is imperative to investigate all possible options to determine the most efficient way of diagnosis, treatment, and prevention. Rapidity, accessibility, and reliability may also represent important objectives for new research. Since the outbreak has already forced businesses across sectors to shut down their activities, many healthcare and technology firms are coming in to fight the pandemic.

In response to the shortfall of intensive care ventilators, start-ups such as Nocca Robotics (incubated at Indian Institute of Technology (IIT)—Kanpur), Aerobiosys Innovations (incubated at IIT Hyderabad), and AgVa Healthcare are working to develop affordable, user-friendly, and portable ventilators that can be deployed even in rural areas of the country. Indian Institute of Science in collaboration with MIISKY Technovation Pvt. Ltd. (April 2020) introduced a device (a smartwatch) that can measure various vital parameters of individuals under isolation on a real-time basis. Linked to a smartphone via Bluetooth, the smartwatch will record the patient's blood oxygen saturation and body temperature in real-time. Experts from the Henry Ford Innovation Institute created COVID care kits filled with different tools that patients may need for disease self-management outside of the hospital. IIT Delhi has also come up with an affordable testing kit to fight against COVID-19. In the meantime, IIT Guwahati is developing robotic units to carry drugs and food to isolation wards to prevent access to healthcare personnel. PerSapien Innovations, New Delhi-based company, has ended up producing a single mask that is capable of protecting the eye, nose, mouth, and ear of COVID-19 patient healthcare providers.

5.4   Non-contact approaches to physiological measurement

Reports from the WHO suggest that one way to contain the disease outbreak is effective screening procedures, as illustrated in the section 5.3.1. This involved identifying individuals with fever, which is a possible symptom of contracting diseases such as SARS or novel coronavirus. SARS-Cov-2 disease (COVID-19) is an infectious disease caused by a coronavirus and patients infected with the virus mostly report mild to moderate respiratory symptoms like fever, tiredness, and dry cough. Some patients may have aches and pains, nasal congestion, runny nose, sore throat, or diarrhea. The main symptom of diseases such as SARS and novel coronavirus is fever onset in the affected individuals. A very common screening approach in public places to separate the coronavirus-infected person is the measurement of their body temperature. Infrared-based thermal scanners are being used to check the skin temperature of individuals in public places. Screening people for possible hyperpyrexia in crowded places such as airports, buses, train terminals, and public places is a complex classification problem. Screening of individuals in public places leads to time-consuming and misclassification when temperature alone is considered.

During this pandemic, the primary duty of the healthcare sector is to identify an individual or group with serious illness, without admitting all symptomatic patients in hospitals or isolating them in quarantine places. Besides, it is important to identify seriously ill patients in the preliminary stage, thereby the point of deterioration where they can be extremely challenging to handle in both pre-clinical and hospital environments can be avoided. Apart from initial temperature screening, if any person is found to be affected, the present testing methods like viral tests and antibody tests are usually practiced. The CDC recommends a COVID-19 test using a nasopharyngeal swab. The technician will insert a special 6-inch cotton swab into both nostrils one by one and move it around for about 15 seconds. It will not hurt, but it might be uncomfortable. This sample is sent to the laboratory for further testing. However, these results yield uncertainties of a person to be having symptomatic or telltale signs. In a few cases, the time duration to receive the testing laboratory results is delayed. The number of COVID tests performed by the government healthcare professionals on daily basis is not proportionate for large population countries like India.

5.4.1 Need for non-contact measurement

Recently, the US Food and Drug Administration (FDA) has insisted the manufacturers of specific FDA-cleared noninvasive, vital sign-measuring devices to expand their use, so that healthcare providers can use them to monitor patients remotely. The devices include those that measure body temperature, respiratory rate, HR, and blood pressure. These are primary signs of concluding that the person is affected by the coronavirus. FDA Principal Deputy Commissioner said, allowing such devices to be used in remote places like home and office can help healthcare providers access information, thereby reducing the need for preliminary tests and also minimizes the risks of spreading the disease to others. Another practical constraint is deploying all the virtual devices like Electrocardiogram (ECG) machines, pulse oximeter, infrared thermometer, and spirometer (respiration measuring device) or multiparameter monitors in public places. That is expensive and requires trained technicians to operate. Sterilization of devices is mandatory before every examination starts, once the data are read for one individual. Considering all the above issues, a method has been proposed here to assess the COVID-19 vital signs (HR, respiratory rate, oxygen saturation (SpO_2), and temperature) from the victims face (Chen et al., 2015; Mori et al., 2016) using thermal and/or visible light scanners (Smilkstein, Buenrostro, Kenyon, Lienemann, & Larson, 2014). This type of non-contact physiological measurement setup would lead to large population screening, particularly in containment zones, airports, malls, industrial estates, educational institutions, and places where more people are permitted (Fig. 5.1).

FIGURE 5.1   Use of non-contact physiological measuring device at various crowded places.

5.4.2   State of the art to prior work

Many research works are found in peer-reviewed articles for non-contact measurement of physiological signals. These inventions are available only in the written form with obtained results.

Ji-Jer Huang, Syu, Cai, and See (2018) developed a capacitive-coupled device embedded on a sofa monitor in the ECG system in a contactless environment. Such a system was able to calculate HR variability parameters. Sun et al. (2017) developed an improved thermal/red, green, and blue (RGB) imaging techniques for a touchless measuring device to estimate HR, respiratory patterns, and skin temperature using Complimentary Metal Oxide Semiconductor- Infrared (CMOS-IR) equipped camera. They experimented in a real-time environment by applying the infrared rays to observe the respiratory and temperature of the subject. The RGB images produced from the CMOS camera were used to predict the HR. Matsunaga, Izumi, Kawaguchi, and Yoshimoto (2016) proposed a microwave-based Doppler sensor to achieve better usability in measuring HR rather than any skin contact sensors or electrodes. Instantaneous HR measurement results aid an individual to manage cardiac-related symptoms and mental stress conditions. Cho et al. (2017) also proposed a webcam-based contactless HR measurement method. Such a system has an HSV + mIIR algorithm for appreciating a robust HR measurement from RGB colored facial images. Tang, Lu, and Liu (2018) presented a versatile measuring practice to monitor ECG signals in unhindered environments by considering the advantage of CNN architectures for skin detection and camera-based remote photoplethysmography method. The advantage of utilizing the CNN model enhances the robust monitoring of

HR, thereby increasing the skin feature extraction in a single step rather than multiple steps like face tracking, detection, and classification.

5.4.3 Proposed approach

The main objective of this work is to identify a person with COVID-19 symptoms in a crowded place without any specialized measuring device. Thermal image scanning cameras are widely used in the measurement of temperature. The color distribution in the scanned images can provide the levels of body temperature. These cameras can only pick up "elevated skin temperatures," not measure the warmth of persons' skin. Henceforth it urges the suspecting individuals to undergo secondary screening, also makes the person a carrier to spread the disease, if the diagnosis is delayed. A high temperature or fever is just one common symptom of the virus. Others include nausea, headaches, fatigue, and loss of taste or smell. It is inferred that not every individual with the virus gets a high temperature and not everyone with a high temperature is infected with the coronavirus. Therefore IR cameras alone will miss infected people with other symptoms or no symptoms at all—known as false negatives. They will also identify people unwell with a fever for other reasons—known as false positives. Considering this, the other signs of COVID-19 are also extracted. As discussed in section 5.2, many works have been carried to measure physiological parameters (like HR, respiration rate, and SpO_2) from thermal imaging. With a walk-through screening, it is possible to extract all the signs from facial thermal images within 30 seconds. The whole process is explained in the diagrammatic approach shown in Fig. 5.2.

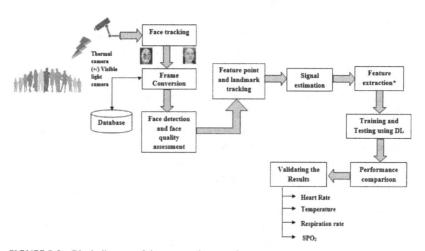

FIGURE 5.2 Block diagram of the proposed approach.

5.4.4 Methodology

Fig. 5.2 shows two different lights (i.e., infrared and visible) to be illumi-nated on a person. The purpose of using two light sources is to make the measurement system cost-effective with greater efficiency. The images of a person are recorded for 30 seconds (1280×720 at 120 fps) under different lighting conditions. The brightness of ambient light can be measured by the lux meter. Persons with different skin colors will be considered for the experiment. Hence, two different databases consisting of thermal and visible light objects are created.

If a frame has more persons, their faces are tracked and identified using facial recognition algorithms. All these objects are converted into frames without affecting the quality. During the recording of subjects, their physio-logical parameters are recorded with a multiparameter monitor to verify the predicted results of our approach. The feature points related to HR, tempera-ture, respiratory rate, and SpO_2 are located in the Region of Interest (RoI) and mostly correlated with various color spaces, RGB, the HSV (hue, satura-tion, and value), and the Lab (lightness, a, and b) color spaces.

The extracted features are given as reference (or) training dataset to the deep learning (DL) model. The purpose of implementing DL for this study is for the consideration of more objects, frames, and training datasets. Machine learning algorithms such as neural networks, support vector machines, and similar kinds of approaches cannot handle high dimensional real-time data. The features are extracted from two different databases created by thermal and visible lighting conditions. All these features are given separately to the DL model. The test data will then be given to the trained model to predict the COVID-19 vital signs. All these predicted results can be statistically cor-related with the actual measurements observed during the recording. These correlation results will be useful to define the overall system performance. The vital sign measurements would be the initial screening parameters before COVID-19.

5.4.5 Preliminary experimental results

Preliminary work has been carried out (Jeyakumar, 2020) to experiment on non-contact measurement of HR from subjects' facial images under the visible light spectrum. A dataset of 160 facial videos each 30 seconds long with the corresponding HR is introduced. The dataset is designed to test the robustness of HR estimation methods. 20 subjects performed four activities (normal with open eyes and closed eyes, running, listening to music) in two lighting setups as mentioned in Table 5.1.

Each activity was captured by the Apple iPhone XS Max mobile camera placed in a tripod with a pixel dimension of 1920×1080. Simultaneously, the corresponding HR was measured using Larsen and Toubro Patient

TABLE 5.1 Subjects selection based on skin tone and lighting condition.

Subjects					Facial videos (per subject)			
Male	Pale skin	5	10	20	Normal with eyes open	1[a]	1[b]	8
	Dark skin	5			Normal with eyes close	1[a]	1[b]	
Female	Pale skin	5	10		After listening to mild music	1[a]	1[b]	
	Dark skin	5			Running (50 stairs)	1[a]	1[b]	

[a]Natural lighting.
[b]Dim lighting.

Monitor (Model No Star Plus-300) with a photoplethysmographic sensor. The range of age of the subjects was from 19 to 29 years, their mean HR was 110 (approx.), and the standard deviation was 25 (approx.). These videos were converted into frames before feeding into the model. The frame rate was set to 0.5 fps. Therefore, each video had been converted into 60 frames approximately, thus obtaining 10,553 images. The participants who participated in this experiment had been given informed consent forms (Fig. 5.3).

These images were fed as input to the system. Very high dimension images (1920 × 1080) are undesirable as they take up extra computation time. Hence all the images were resized to a dimension of 64 × 64. Image resizing was done using the CV2 package in python. The analysis was made to find the correlation between facial features and HR. The following inference was made after the analysis of the correlation between facial images and HR.

• There exists a correlation between the mean intensity of the green channel and the HR.
• As the HR increases, the mean intensity of the green channel decreases.
• The wavelengths concerning green colors have maximum absorptivity, and evident that rise in blood volume leads to less absorptivity by red color (Fig. 5.4).

5.4.5.1 Face detection and region of interest selection

The face was detected from the frame by a face detector. The forehead region was isolated as the RoI from the whole frame by using the Viola − Jones algorithm (Haas-cascade classifier) (Fig. 5.5).

The proposed system consists of two parts. The videos were split into frames and were separated into three parts—training set, testing set, validation set, and

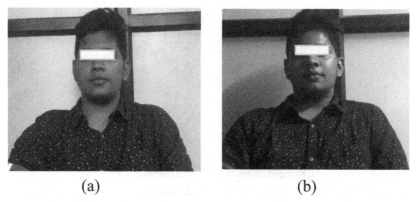

(a)   (b)

FIGURE 5.3 (A) Facial image captured under ambient lighting condition. (B) Facial image captured under dull light after running about 100 meters.

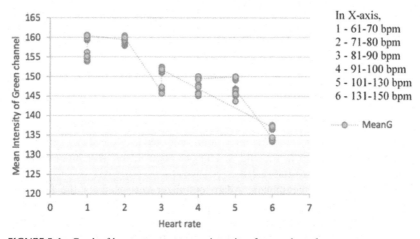

In X-axis,
1 - 61-70 bpm
2 - 71-80 bpm
3 - 81-90 bpm
4 - 91-100 bpm
5 - 101-130 bpm
6 - 131-150 bpm

····◉···· MeanG

FIGURE 5.4 Graph of heart rate versus mean intensity of green channel.

annotated. In the first part, the video was split into frames that were annotated/labeled. These frames were split into training, testing, and validation sets in the proportion of 80:15:5. These frames were fed into the proposed system as input along with the respective HRs. The output was a numerical number, which represented the HR. The correlation between the predicted HR from the system and the actual HR value was calculated to determine the system's accuracy. To feed the data into the CNN model, the pretrained models such as Alexnet and Lenet were initially considered. Due to the overfitting issue, these models provided a high misclassification rate during the testing phase. As an alternative to these issues, a smaller version of Visual Geometry Group (VGG) network with fewer weight layers was considered to make the training period less (Fig. 5.6).

FIGURE 5.5 Face detection and region of interest (RoI) selection.

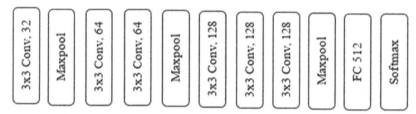

FIGURE 5.6 A smaller version of VGG net.

The model built had six layers in total, not counting the max pool layers and the Softmax layer at the end. It had one input layer, five hidden layers and one output layer (Table 5.2).

To optimize the parameters, Adam optimizer was used for the first and second moments of the gradient for updating the learning rate adapting to each training image. The above VGG model was trained with the training images with the hyperparameter tunable optimization function. During the training, the maximum validation accuracy of 74.66% and minimal loss of 10.1% at epoch 50 was obtained. A sample of images was given during the testing phase and the trained model provides the predicted values of the HR. The accuracy of the trained system was calculated from Eqs. 5.1 and 5.2.

$$\text{Error } (\%) = \frac{(\text{Obtained HR value} \sim \text{Actual HR Value})}{\text{Obtained HR value}} * 100 \quad (5.1)$$

$$\text{Accuracy} = 100 - \text{Error } (\%) \quad (5.2)$$

The CNN was implemented by python with Keras package using NVIDIA GeForce GTX 1060 GPU. The predicted results of the model are shown in Fig. 5.7.

For the above test images, the accuracy rate was 97.5%, 95.96%, and 95.45%, respectively. The overall mean accuracy for the test images was

TABLE 5.2 Hyper parameter values.

Number of filters	32, 64, 128
Size of filter	3 × 3
Max pooling shape	2 × 2
Stride	1
Padding	The number on the edge
Learning rate	0.01
Number of epochs	50
Batch size	64
Activation function	Relu, Softmax
Hidden layers	5
Dropout	0.25, 0.5

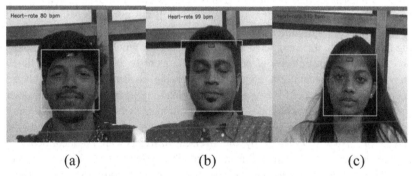

     (a)   (b)   (c)

FIGURE 5.7   HR values of A O.value—80 Bpm; A.value—82 Bpm B O.value—99 Bpm; A. value—95 Bpm C O.value—110 Bpm; A.value—105 Bpm.

92.32%. Still, there is scope to improve the system's overall efficiency by upgrading the processor and considering more data to train the model. In this preliminary study, HR alone was considered from the facial attributes. However, other vital parameters like respiration rate, SpO2, and temperature can also be measured from the approach proposed here.

5.5   Conclusion

The number of positive coronavirus cases continues to increase at an unprecedented rate globally which urges the doctors, nurses and other medical professionals to work round the clock. This affects their physical and psychological condition badly. Hence a measuring system to screen the people in a large scale is the need of the hour.

Temperature measurement is a part of the evaluation to decide whether a person has a high temperature potentially caused by an infection with COVID-19. Using various instruments for temperature measurement, such as oral thermometers, involves physical contact that can increase the risk of infection spread. Thermal imaging systems and non-contact infrared thermometers, instruments for non-contact temperature measurement are used to measure an individual's temperature. These non-contact temperature evaluation devices can quickly measure and display a reading of temperature, so that a large number of people can be individually assessed at entry points. Such instruments have many benefits but they must be used properly to get correct readings. The elevated temperature alone cannot be taken into consideration to confirm COVID-19, further assessments of other vital parameters are needed to evaluate in non-contact environment.

Certain proven studies and results are available in the literature stating that the various physiological parameters can be determined from the facial images. Most of the COVID-19 tests are being done due to a rise in body temperature alone. This culminates in most of the test results being declared as asymptomatic cases of COVID-19 if the temperature is not measured. If this approach is implemented in the current scenario, the number of true positive cases will increase. By adopting this approach, the mass screening of individuals or groups in industrial, home, and educational institutions is possible. Daily increase of positive COVID-19 cases in green containment zones is another big challenge to the healthcare department. This happens because of migrants moving from the red zone to other low case load containment zones. If this measurement system is installed at entry points such as toll gates and immigrant zones of airports and harbors, then the spread of SARS COVID-19 by the overseas travelers can be identified and the travelers can be isolated and quarantined. The systems would be beneficial to the persons who are undergoing self-quarantine at their home. Moreover, these systems would be an assistive device to assess the vital signs of a vulnerable group. These systems can be incorporated in any surveillance systems like CCTV and Drone. These would be more beneficial to healthcare professionals. Such a system is useful to find cases of infectious diseases and for the continuous health monitoring of elderly people, sportspersons, and young children, and also for remote monitoring, follow-up and communication with the patients without human contact and risk of infection to the healthcare workers.

Acknowledgment

The authors would like to thank Ruba M, Kousika V, Viveka S, and Gurucharan M K who contributed to preliminary experimental results of non-contact measurement of heart rate from facial images.

References

Bhagat, R. B., Reshmi, R. S., Sahoo, H., Roy, A. K., & Govil, D. (2020). April 14 *The COVID-19, migration, and livelihood in India*. Mumbai: International Institute for Population Sciences.

Bhatnagar, V., Poonia, R. C., Nagar, P., Kumar, S., Singh, V., Raja, L., & Dass, P. (2020). Descriptive analysis of COVID-19 patients in the context of India. *Journal of Interdisciplinary Mathematics*, 1−16. Available from https://doi.org/10.1080/09720502.2020.1761635.

Carter, C., Aedy, H., & Notter, J. (2020). COVID-19 disease: Assessment of a critically ill patient. *Clinics in Integrated Care*, *1*, 10001.

Chen, D. Y., Wang, J. J., Lin, K. Y., Chang, H. H., Wu, H. K., Chen, Y. S., & Lee, S. Y. (2015). Image sensor-based heart rate evaluation from face reflectance using Hilbert-Huang transform. *IEEE Sensors Journal*, *1*, 618.

Cho, D., Ham, J., Oh, J., Park, J., Kim, S., Lee, N.-K., & Lee, B. (2017). Detection of stress levels from biosignals measured in virtual reality environments using a kernel-based extreme learning machine. *Sensors (Basel)*, *17*(10), 2435.

Jeyakumar, V. (2020). *Non-Contact Pulse Rate Measurement using Facial Videos*. 2020 IEEE International Conference on Advances and Developments in Electrical and Electronics Engineering (ICADEE) (pp. 1−6). USA: IEEE. Available from https://doi.org/10.1109/ICADEE51157.2020.9368944.

Ji-Jer Huang, H., Syu, H.-Y., Cai, Z.-L., & See, A. R. (2018). Development of a long-term dynamic blood pressure monitoring system using the cuff-less method and pulse transit time. *Measurement.*, *124*, 309−317.

Kumari, R., Kumar, S., Poonia, R. C., Singh, V., Raja, L., Bhatnagar, V., & Agarwal, P. (2020). Analysis and predictions of spread, recovery, and death caused by COVID-19 in India. *Big Data Mining and Analytics*. Available from https://doi.org/10.26599/BDMA.2020.9020013.

Longb, C., Xuc, H., Shend, Q., Zhang, X., Fana, B., Wanga, C., . . . Honglu, L. (2020). Diagnosis of the coronavirus disease (COVID-19): rRT-PCR or C.T.? *European Journal of Radiology*, *126*, 108961.

Loske, D. (2020). The impact of COVID-19 on transport volume and freight capacity dynamics: An empirical analysis in German food retail logistics. *Transportation Research Interdisciplinary Perspectives*, *6*, 100165.

Matsunaga, D., Izumi, S., Kawaguchi H., & Yoshimoto, M. (2016). Non-contact instantaneous heart rate monitoring using microwave Doppler sensor and time-frequency domain analysis. 2016 IEEE 16th International Conference on Bioinformatics and Bioengineering (BIBE). pp. 172−175.

Mori, T., Uchida, D., Sakata, M., Oya, T., Nakata, Y., Maeda, K., . . . and Inomata, A. (2016). Continuous real-time heart rate monitoring from face images. Proceedings of the 9th International Joint Conference on Biomedical Engineering Systems and Technologies: BIOSIGNALS, (BIOSTEC 2016). pp. 52−56.

Nicolaa, M., Alsafib, Z., Sohrabic, C., Kerwand, A., Al-Jabird, A., Iosifidisc, C., . . . Aghaf, R. (2020). The socio-economic implications of the coronavirus pandemic (COVID-19): A review. *International Journal of Surgery* (78), 185−193.

Singh, V., Poonia, R. C., Kumar, S., Dass, P., Agarwal, P., Bhatnagar, V., & Raja, L., L. (2020). Prediction of COVID-19 coronavirus pandemic based on time series data using support vector machine. *Journal of Discrete Mathematical Sciences & Cryptography*. Available from https://doi.org/10.1080/09720529.2020.1784525.

Smilkstein, T., Buenrostro, M., Kenyon, A., Lienemann, M., & Larson, G. (2014). Heart rate monitoring using Kinect and color amplification. In Healthcare Innovation Conference (HIC), IEEE. pp. 60−62.

Sun, G., Nakayama, Y., Dagdanpurev, S., Abe, S., Nishimura, H., Kirimoto, T., & Matsui, T. (2017). Remote sensing of multiple vital signs using a CMOS camera-equipped infrared thermography system and its clinical application in rapidly screening patients with suspected infectious diseases. *International Journal of Infectious Diseases*, *55*, 113−117.

Tang, C. X., Lu, J., & Liu, J. (2018). Non-contact heart rate monitoring by combining convolutional neural network skin detection and remote photoplethysmography via a low-cost camera. The IEEE Conference on Computer Vision and Pattern Recognition (CVPR) Workshops. pp. 1309−1315.

World Health Organization (WHO), (2020) The Unity Studies: WHO Early Investigations Protocols. https://www.who.int/emergencies/diseases/novel-coronavirus-2019/technical-guidance/early-investigations Accessed October 31, 2020.

Zhu, J., Ji, P., Pang, J., Zhong, Z., Li, H., He, C., . . . Zhao, C. (2020). Clinical characteristics of 3062 COVID-19 patients: A *meta*-analysis. *Journal of Medical Virology* (92), 1902−1914.

Chapter 6

Evolving uncertainty in healthcare service interactions during COVID-19: Artificial Intelligence - a threat or support to value cocreation?

Sumit Saxena and Amritesh

Department of Humanities & Social Sciences, Indian Institute of Technology, Ropar, India

6.1 Introduction

Healthcare is highly complex, yet the most fertile field for service research (Berry & Bendapudi, 2007). The transforming nature of the healthcare environment is characterized by a change in the working style, adoption of modern practices, advanced technology, empowering patients, and an increasing number of stakeholders. Adding to this list, information asymmetry and healthcare services' credence make it more complicated and turbulent (Bloom, Standing, & Lloyd, 2008). The technology revolution happening since the past few decades has been transforming healthcare services (Duplaga, Zieliński, & Ingram, 2004). Earlier, the healthcare consumer simply used to play a passive role because of a low level of access to health information. They are now empowered through digital information and interaction services and have become more demanding. This transformation has increased customer expectations and created tensions in the service interaction environment. As a result, the overall situation challenges the service provider's way of designing and offering services. To reduce this tension in healthcare service interactions, researchers have proposed a cocreation approach where both the actors, for example, service provider and service recipient, cocreate the value equally (Barile, Saviano, & Polese, 2014). In other words, using the cocreation approach, the service provider (i.e., the doctor) and the healthcare consumer (i.e., the patient) cocreate value for each other to realize their real wellbeing (McColl-Kennedy, Vargo, Dagger,

Cyber-Physical Systems. DOI: https://doi.org/10.1016/B978-0-12-824557-6.00014-5

Sweeney, & Kasteren, 2012). McColl-Kennedy et al. (2012) have beautifully defined cocreation within healthcare as *benefit realized from integration of resources through activities and interactions with collaborators in the customer's service network*. Here activity means any "doing" by either patients or doctors/service providers. This cocreation phenomenon is emerged from a well popular logic within service marketing, the service dominant logic (SDL) (Vargo & Lusch, 2004). SDL states that service is the basis of exchange in any economy where each actor acts as a resource integrator and creates value for each other. This "value for each other" results in "cocreated value" and the process is coined as "value cocreation" (Vargo & Lusch, 2004). Resource integration involves optimally combining several tangible and intangible resources from various sources of service providers and consumers. For example, consumers may use their resource (like their knowledge and skills) and combine it with the public resources (like the information available on the internet), along with the resources provided by the hospitals (e.g., an expert consultation, diagnostic and treatment facilities) to cocreate health outcomes in the best possible way.

Unfortunately, healthcare value cocreation (VCC) interaction spaces get disrupted during the pandemic's difficult time like COVID-19. Increasing health risks and a lack of appropriate response desired from the healthcare systems are giving rise to fear and anxiety worldwide. Several healthcare uncertainties emerge radically during the pandemic and hinder the effective practices of VCC (Han, Klein, & Arora, 2011). For instance, doctors may avoid direct interaction with the patients and prefer to spend less time with them than the legitimate normal due to the fear of chronic communicable infections. The mutual "trust" that is essentially required between a doctor and a patient for a successful cocreation in a service encounter is severely affected by the pandemic. Healthcare consumers often start doubting the service provider's (doctors) honesty, fairness, and communication intention. Lack of procedural justice, conflict of interest, lack of empathetic communication, and ethical perception toward service pose challenges to cocreation practices. Some of these issues, which were already hidden in the healthcare service system, have now been exaggerated due to pandemic situations.

In parallel to the above-discussed background, recently, the experts around the world are advocating to deploy Artificial Intelligence (AI) as a magical tool to solve all the uncertainties of healthcare services during COVID-19 (Nguyen, 2020). Nguyen (2020) completed a comprehensive survey on AI use during COVID-19 and finds the possibility of its application into many areas such as medical image processing (through deep learning methods), pandemic modeling (by using data science), medical devices (by AI and the Internet of Things), text mining and natural language processing (NLP), and in computational biology and medicine. In the VCC context, Robinson et al. (2020) proposed the AI-empowered service encounter framework, where authors deliberate over AI-enabled interactions in a dyadic service exchange. Much of the growing research to date primarily focuses only on applying AI in a broader service system.

The larger question that needs attention is how AI influences the "cocreation process" within the healthcare service system. A similar question was earlier raised by Kaartemo and Helkkula's study, who reviewed the role of AI & robots in VCC (Kaartemo & Helkkula, 2018). Augmenting further, Čaić, Odekerken-Schröder, and Mahr (2018) observe that AI (such as in the form of service robots) may also result in value codestruction and may spoil the overall service experience, especially in the context to healthcare services. Thus, it becomes imperative to critically explore AI's role, that is, observe the negative side of AI from a balanced perspective. Based on the above discussions, the primary question posed in this work is, "Is the present AI system capable enough to solve the uncertainties (due to pandemic) underlying cocreative service interactions and support the healthcare VCC?" This chapter focuses on conceptualizing this question in the light of interdisciplinary studies from services marketing, healthcare and technology literature. We try to answer some of the interrelated questions like how AI can resolve the issues of trust, social support, and negative ethical perceptions that emerged due to the pandemic. Without questioning the technical competence of AI, it is critical to question the applications of AI guided by a phenomenological view.

First of all, we identify the natural uncertainties emerging from COVID-19 in cocreative healthcare service interactions and then explore AI's role as an advocated solution. This work is a conceptual study presenting an exciting viewpoint about the role of "AI" in combating COVID-19, which intersects with a niche theoretical understanding of VCC in healthcare services. We balance the literature by highlighting the lesser focused side of AI, that is, adverse effects of AI on VCC, which is likely to have undesired outcomes. Our discussion will follow the line of thoughts of Čaić et al. (2018), LaRosa and Danks (2018), and Xu, Shieh, van Esch, and Ling (2020), who present the second side of the coin, that is, reflecting the spill-over effect of AI in healthcare services. Overall, the contribution of this work is threefold. First, it opens up discussing possible value codestruction in AI-enabled services, especially in healthcare. Second, this study helps to understand the dynamics of healthcare uncertainties evolved during pandemic like COVID-19. Third, it elaborates the AI's potential to combat service uncertainty from social science perspectives by adopting the consumer's view instead of AI's technical competence. Further, the complete study is organized as follows: First, discussing the SDL in marketing; second, talking about service interactions and cocreated wellbeing; third, presenting uncertainty due to pandemic; fourth, highlighting uncertainty in healthcare; fifth, elaborating the emerging role of AI; sixth, focusing upon AI combating uncertainty and supporting VCC in healthcare interactions; seventh, discussing the spill-over effect of AI; and eighth, summarizing overall work through conclusion along with limitations and future research directions.

6.2 Service dominant logic in marketing

Marketing's true philosophy is rooted in the notion that *customer is the king* and company should offer *relevant value to their customers* (Kotler, 1994). However, the way value is offered changed with time. Earlier, where the value was offered through the product has now changed to value through service (Sheth & Uslay, 2007; Vargo & Lusch, 2004). Pioneers of SDL, Ramaswamy (Prahalad & Ramaswamy, 2004) and Vargo (Vargo & Lusch, 2004) agree that even a tangible product offered to customers perform services to fulfill the customer's need. For example, a car is not seen as an automobile product but as a comfortable transport service coupled with luxury, esthetics, status, power, and even self-esteem for a few customers. This notion gets stronger with the wide adoption of the premises of SDL. This logic simply believes that every market transaction is concerned with service, and service is the primary basis of exchange in the economy. It implies that the customer who enjoys the logistic service of the purchased car results from an engineering service that turns raw material into finished automobile products. This idea replaces the traditional concept of exchange of goods to the modern concept of exchange of services (Lusch, Vargo, & O'brien, 2007). Since then, several premises of SDL have augmented the service approach of marketing. One of the important premises directly relevant in context to this study (talking about value in service interactions) is *The customer is always a cocreator of value*. This implies that a firm cannot deliver value but can only offer a value proposition. It is also implied from here that customer plays an active role in the service interaction. Another essential SDL premise relevant for our study is, *Value is always uniquely and phenomenologically determined by the beneficiary*. This implies that the service provider's offered value would be worthful for the beneficiary only when it is infixed with his active involvement. The customer finally realizes the value in any service interaction using his operant resources like knowledge and skills (Vargo & Lusch, 2004, 2016).

Drawing from this SDL, it could be said that every actor (while consuming the services) creates value not only for himself but also for other actors in the service network. Relating this thought (of creating value for other actors in the network) in healthcare, it is observed that healthcare actors (doctor, patient, paramedical staff, information and communications technology team, back-office personnel) create value for each other using both operant and operand resources (Hau, 2019). Although, earlier SDL studies focus more on operant resources (customer knowledge, skills, compliance, psychological states) contrary to operand resources (technology) in healthcare VCC (Robertson, Polonsky, & McQuilken, 2014). Considering this opportunity, the current study focuses on the possible role of AI (an emerging technology) in healthcare VCC (rooted in service-dominant healthcare interactions). In addition, the context of COVID-19 demands more investigation into the above issue as the pandemic poses several challenges (physical, psychological, social, and technological) to effective cocreation of value (Finsterwalder & Kuppelwieser, 2020).

6.3   Service interactions and cocreated wellbeing

The service interactions in any industry are oriented toward an outcome, which may be "profit for service provider" or "value for money" for the customer. However, such outcomes may be contradictory, and achieving the real harmonious outcome is challenging. Here, harmony means the outcome that emerged in service interaction should be considered equally valuable by both the provider and the consumer. This challenge gets further critical in healthcare services because of its complex nature (Verleye et al., 2017). To better understand, let us take an example. A patient may be actively participating, such as understanding the health issues in more detail by referring to alternative (though not always credible) information sources, using specific drugs without consultation, etc. to realize self-empowerment or self-control value creation processes. This may endanger the doctor's role to make patients strictly compliant with the suggested line of treatment. Literature asserts that there are several reasons for such challenges in healthcare service interactions. Some of these are the credence trait of healthcare services (Darby & Karni, 1973), information asymmetry in healthcare (Barile et al., 2014), and power imbalance in healthcare (Verleye et al., 2017).

The role of cocreation in healthcare services becomes vital to addresses these challenges. In healthcare service encounters, whenever one actor perceives other actors to be participating, his cocreation efforts are enhanced, which results in a synergistic outcome that is mutually satisfying in nature (Tari Kasnakoglu, 2016). Several other studies discuss VCC at a dyadic or systemic level in healthcare services (Janamian, Crossland, & Wells, 2016; Kim, 2019; Osei-Frimpong, Wilson, & Owusu-Frimpong, 2015; Virlée, Hammedi, & van Riel, 2020). More recently, Sharma, Conduit, and Hill (2017) identify that healthcare consumers derive different types of personal wellbeing (hedonic and eudemonic wellbeing) from cocreation activities. Extending this cocreation and wellbeing dynamics, Chen et al. (2020) explore the cocreation of wellbeing itself, as wellbeing is the critical outcome expected by all the actors in the healthcare services. It may be said that within healthcare service encounters, doctors try to enhance the patient's medical knowledge so that he can participate, realizing his sense of achievement or eudemonic wellbeing. This wellbeing of the patient fosters a doctor's wellbeing because a doctor can feel a sense of ownership for the patient's improved condition. Thus it is implied that SDL-rooted cocreative service interaction is directly associated with cocreated wellbeing in healthcare.

6.4 Uncertainty due to pandemic

Any kind of pandemic often exaggerates the typical complexity or uncertainty inherent in society. Before moving further, it is necessary to understand that uncertainty and risk are two different concepts. Interestingly, Sharma, Leung, Kingshott, Davcik, and Cardinali (2020) differentiate between risk and uncertainty, elaborating that risks can be evaluated up to a certain extent compared to uncertainty, which is difficult to comprehend in advance. This study will mainly talk about uncertainty only. Elaborating over the nature of uncertainties, MacPhail (2010) categorizes uncertainty as scientific, situational, and strategic uncertainty. Liang, Laosethakul, Lloyd, and Xue (2005) observe that uncertainty emerges either from an actor's opportunistic behavior or due to information asymmetry in any relationship. Shiu, Walsh, Hassan, and Shaw (2011) proposed a multidimensional view of uncertainty, that is, knowledge uncertainty, choice uncertainty, and evaluation uncertainty. Karlsen and Kruke (2018) talk about the uncertainty of the source, evolution, and solutions to the pandemic.

Ultimately, based on a selective review of studies on uncertainty, the critical uncertainties or issues that emerged during a pandemic are identified. For details, refer to Table 6.1.

6.5 Uncertainty in healthcare

Uncertainty is inherent in healthcare services since its inception because there is a significant knowledge gap among focal actors (doctor and patient) in this industry. However, this information asymmetry is reduced to a certain extent, with the advent of ICT (Information & communication technologies) supporting patients' health literacy (D'Cruz & Kini, 2007). Still, there are many other uncertainties as per Han et al. (2011), visible in healthcare service consumption, listed below:

- *Ambiguity: regarding patient's evaluation of their state of illness.*
- *Credence trait of services: due to which consumer could not ascertain the actual quality of services even after consumption.*
- *Unpredictability: due to a change in the patient's condition during the complete course of treatment.*
- *Lay epistemology: patient's subjective interpretation of available information.*
- *Existential issue: influence of patient's condition on his personal life* (Han et al., 2011).

It is evident that healthcare services are filled with uncertainties, even in regular times. For example, if a person is diagnosed with cancer, he may start experiencing anxiety/fear or loss of control over life. All this creates uncertainty in the patient's mind and affects his quality of life or overall

TABLE 6.1 Brief about key uncertainties identified in earlier studies, along with primary focus and COVID-19 issues discussed in the study.

S. no.	Study reference	Key uncertainty identified in the study	Main focus of study	COVID-19 issues discussed or implied in the study
1	Sharma et al. (2020)	Economic uncertainty, political uncertainty, cultural uncertainty, technological uncertainty, behavioral uncertainty, demand uncertainty, disruptions in supply chain	Focus on an extensive review of international business literature to find the key uncertainties, their antecedents, and outcomes	The study especially highlights the "discontinuous uncertainty" representing the COVID-19 associated uncertainty, thereby suggesting relevant uncertainty management strategies.
2	Shiu et al. (2011)	Knowledge uncertainty, choice uncertainty, evaluation uncertainty	Focus on understanding uncertainty through a multidimensional lens and offering a new theoretical model within consumer behavior	The theoretical framework of the study proposes a certain dimension (knowledge uncertainty, ambiguity, and credibility) that rightly fits into COVID-19 circumstances.
3	Karlsen and Kruke (2018)	Uncertainty related to the source of the problem, uncertainty related to the problem evolution, uncertainty about the possible solutions to the problem, increasing fear and stigma in society	Focus on understanding the impact of uncertainty as an enabler of "non-decision-making," especially on actions that could have slowed the rate of pandemic escalation	It mainly discusses how pandemic-led uncertainty is socially constructed. Although it talks about the Ebola virus, it is equally applicable here because of its high similarity with COVID-19.

(Continued)

TABLE 6.1 (Continued)

S. no.	Study reference	Key uncertainty identified in the study	Main focus of study	COVID-19 issues discussed or implied in the study
4	Quintal, Lee, and Soutar (2010)	Financial uncertainty, physical uncertainty, uncertainty about performance, psychological uncertainty, uncertainty of leisure in life, uncertainty of travel and loss of emotional support	Focus on observing the influence of perceived uncertainty and risk on consumers' decision-making using theory of planned behavior	It talks explicitly about "uncertainty avoidance" (which measures the extent to which people get scared by unknown situations) that has a direct implication for the COVID-19 uncertainty study.
5	Brashers (2001)	Uncertainty about self or other's communication skills, uncertainty about affective state of individual, uncertainties about own beliefs, behavior, and values, revival uncertainty (experienced in chronic illness), micro-interactional uncertainty, temporal uncertainty, and the social uncertainty	Focus on understanding the uncertainty and the underlying communication processes, thereby explaining the overall uncertainty management	It primarily discusses issues (uncertainty experience, emotion in uncertainty management, and the psychological response to uncertainty), which has implications for COVID-19 uncertainty studies.
6	Usher, Durkin, and Bhullar (2020)	Intolerance of uncertainty, increasing fear, decreasing sense of meaning or purpose in life, feeling of loss of	Focus on evaluating the pandemic-led uncertainties/panic (adopting the psychosocial viewpoint) and	It elaborates upon the several forms of psychosocial uncertainties emerged due to pandemic (COVID-19) and

(Continued)

TABLE 6.1 (Continued)

S. no.	Study reference	Key uncertainty identified in the study	Main focus of study	COVID-19 issues discussed or implied in the study
		control, growing inequality in society, uncertainty of social structures for moral support	offering practical solutions in context to mental health	leading toward an unstable or fearful environment.
7	Zacher and Rudolph (2020)	Uncertainty about life satisfactions, uncertainty about morality or ethics in society, health-related worries, uncertainty about job security, work–family conflict, social discrimination, uncertainty about the level of controllability and psychological energy	Focus on elucidating the change in subjective wellbeing of individuals during the early stage of COVID-19	The study identifies that different stress appraisals and pandemic (like COVID-19) coping strategies play a limited role in changing individuals' subjective wellbeing.

wellbeing. Under such circumstances, if patients receive appropriate support from healthcare interactions, they may experience a lesser disruption in wellbeing (Arora, 2003). Here, supportive healthcare interactions could be reflected through key activities like active listening of the patient, estimating patient's information curiosity, empowering patients through shared decision making, offering emotional support to patients and his family, offering psychosocial support postconsultation, offering spiritual support in critical stage, enhancing esthetics in the hospital setting, and creating an environment of trust based on interpersonal relationships. All this reflects that service providers can add value to the patient's efforts toward their wellbeing (Sweeney, Danaher, & McColl-Kennedy, 2015). Healthy service interactions can result in wellbeing cocreation, and this is also evident in the emerging literature on transformative service research (Kuppelwieser & Finsterwalder, 2016). Unfortunately, cocreation's fundamental dynamics get disturbed during the pandemic as it directly hurts the healthcare service interactions. It implies that there is a two-way attack on the patient. One is the increase in their anxiety level or perceived uncertainty, and second is the decline in the quality of healthcare interactions.

6.5.1 Impact of pandemic-led uncertainty on a patient's mind

During a pandemic situation, healthcare consumers often experience a state of knowledge uncertainty because of the rumors or contradictory information in the environment. Pandemic imbibes a feeling of insecurity, anxiety, fear, and culpability (Rubin & Wessely, 2020; Usher et al., 2020). Gopalan and Misra (2020) reported deteriorated social relationships and a lack of emotional support due to the pandemic. Wagner-Egger et al. (2011) asserted that during a pandemic, not only just individuals but also the collectives (like whole nation or society) are worse affected. They explore the lay perception of collectives during pandemic (H1N1 outbreak) and frame them as heroes (doctor and researchers), victims (less developed societies), and villains (pharmaceutical firms and media). Adding further, studies assert that pandemic may influence normal patients and patients suffering from chronic noncommunicable diseases (NCDs). Actually, during a pandemic, NCD patients may develop an inactive or sedentary lifestyle (because of lockdown), resulting in their poor physical and mental health.

6.5.2 Impact of pandemic-led uncertainty on service interactions

It is evident in studies that crucial service providers, that is, doctors, may show an unwillingness to continue their work probably due to concern for their family members' health or own health (Ives et al., 2009). Due to government obligation in many countries, several doctors could not discontinue their work despite having a low willingness to treat their patients. This affects their engagement level and quality of services offered by them. Few healthcare professionals, especially nursing staff or paramedical persons, often show unwillingness for work due to nonpersonal reasons like transport problems or lack of trust in management (Ives et al., 2009). Another issue in a pandemic is the lack of social support in healthcare service consumption. Due to fear of a pandemic, patient's relatives or friends may not accompany them while visiting the healthcare facility. Thus consumers may lose a critical resource required for the service encounter, that is, a social resource (Lee, Ozanne, & Hill, 1999). This directly affects consumer VCC, as social support is one of the critical antecedents for healthcare VCC (Hau, 2019). Adding to this list of uncertainty in the pandemic, another issue is the lack of reciprocity or governance support to the healthcare system, directly affecting doctors and staff. Whenever doctors and patients perceive low reciprocity (in terms of exchanging information by government bodies, financial, and legal support), they may feel less enthusiastic in their participation in cocreation activities. This affects their participation at different healthcare ecosystem levels and thus indirectly hurt each other's wellbeing. This also represents a link between macro, meso, and micro levels of cocreation between actors in the healthcare ecosystem (Frow, McColl-Kennedy, & Payne, 2016). Lastly, pandemic often hurts interpersonal trust and consumer ethical perceptions. Both trust and ethics are an essential element of the successful service encounter. Healthcare studies confirm that trust plays an essential role in the doctor–patient service relationship. Patients who perceive high interpersonal trust in healthcare interactions often experience high responsiveness and better quality of services. This interpersonal trust is accessed based on honesty, fairness, and communication (Topp & Chipukuma, 2016). However, there is a significant decline in consumer trust during a pandemic, both institutional and interpersonal (Esaiasson, Sohlberg, Ghersetti, & Johansson, 2020). Thus, it is implied that during a pandemic, the patient may feel more skeptical or doubt the doctor's communication, honesty, and fairness while interacting for cocreation.

6.6   The emerging role of Artificial Intelligence

AI could be understood as machines performing human-like functions and trying to imitate human intelligence in the simplest form. AI has already served society in multiple fields like education, defense, agriculture, manufacturing, finance, and healthcare (Dwivedi et al., 2019). It has emerged rapidly in the last decade, solving complexities in our day- to-day activities. Looking deeper into AI and consumer services, it is evident that AI is embedded in today's service encounters and plays a positive role in consumer service consumption. Here, service encounters could be understood as direct contact (preferably face to face) between the customer and the service provider. Within this encounter, AI plays a vital role at different touch points like assisting frontline service personnel (back office), emerged as a tool to interact with customers (robot in services), and directly offering service to customers (self-service technology). Such roles are labeled as AI supporter, AI-augmented, and AI performer (Ostrom, Fotheringham, & Bitner, 2019). Understanding with an example, like Chatbot (a popular form of AI), has a remarkable capacity to solve customer service issues and also update itself (means self-learn) based on its experience of solving similar issues (Xu et al., 2020). It is possible due to advanced deep learning and NLP trait of AI (Kaplan & Haenlein, 2019). Elaborating more about AI's role in services, Huang and Rust (2018) present an ordered development of AI based on human traits required for effective services. It mainly proposes four types of intelligence that AI has developed over time, that is, mechanical intelligence, analytical intelligence, intuitive intelligence, and empathetic intelligence.

Discussing within healthcare, AI is crucial for patient disease diagnosis, electronic health records (EHR), health insurance claim, patient self-service technology, clinical trials, drug compliance, postsurgery recovery, palliative care, patient empowerment practices, and significant physician—patient interaction (Combi, 2017; Kahn, 2017; Khanna, Sattar, & Hansen, 2013; Thesmar et al., 2019; Zandi, Reis, Vayena, & Goodman, 2019). Apart from the above-discussed importance of AI in standard time, AI's role is highly influential and fruitful during a pandemic. Talking specifically about COVID-19, Nguyen (2020) explores in detail how AI is playing a vital role in the fight against the pandemic and its additional potential that can be harnessed in the future. On the same line, Mahomed (2020) explores how AI can empower the healthcare system during pandemic. Exploring deeper, it is observed that scientists are using emerging AI techniques for combating COVID-19 (Vafea et al., 2020). For example, machine learning models are designed based on genomic signatures to identify the COVID-19 viral sequence; use of "Stereographic Brownian Diffusion Epidemiology Model" for analyzing the spread of the virus; using "Unmanned Aerial Vehicles" for thermal scanning; 3D printing for producing ventilator splitters; "Deep Convolutional Neural Network" for medical imaging and diagnosis; and "Hybrid Wavelet-autoregressive integrated moving average model" for early forecasting of COVID-19 cases (Lalmuanawma, Hussain, and Chhakchhuak, 2020; Vafea et al., 2020). Recently, related researches (focusing on the emerging role of advanced mathematical modeling, big data mining, and emerging AI against COVID-19) have initiated within developing countries (like India and South Africa) as well (Bhatnagar et al., 2020; Kumari et al., 2020; Priyanka Harjule, Agarwal, & Poonia, 2020; Singh et al., 2020).

6.7 AI combating uncertainty and supporting value cocreation in healthcare interactions

We first discuss how AI can be used for combating uncertainty in cocreative service interactions evolved due to the pandemic. As already discussed earlier, a pandemic can affect healthcare services in multiple ways like doctor's unwillingness to treat their patients due to concern for their own safety, low level of service provider's engagement in services, emerging lack of interpersonal trust, lack of social support, or low level of social resource within service encounters due to stigma, lack of reciprocity, poor governance support, negative ethical perceptions, low level of perceived honesty, perceived fairness and perceived communication ability.

We will first look into these aspects and try to find theoretically (based on literature) if AI can resolve them directly or indirectly:

- AI can partially resolve the issue of doctor's unwillingness to treat their patients. AI-based self-diagnostic tools and medical decision support systems can facilitate the virtual encounter between doctor and patient, reducing physical touch length in service encounters. Thus the risk posed to a healthcare professional is reduced (Reddy, Fox, & Purohit, 2019). This reduced risk may result in the reduced unwillingness of doctors to treat their patients during a pandemic.

- Regarding the low level of provider engagement in service offerings, AI can work indirectly in this context. Organizations can measure their employees' social media activities using predictive analytics, which helps monitor their overall engagement level (King, Tonidandel, Cortina, & Fink, 2015). This monitoring helps in managing employee engagement in the workplace. This employee engagement can foster customer engagement in services. Thus AI may indirectly influence the engagement level of service employees and customers as well.

- Lack of interpersonal trust in services is another essential concern that emerged during the pandemic. AI can partially resolve this issue as consumers report a high level of cognitive trust within a virtual environment, that is, while interacting with AI-based Chatbots (Kanawattanachai & Yoo, 2002). Cognitive trust is based on the rational judgment of the actor's competence and knowledge (Butler, 1991), which is generally very high among virtual actors than the real actor. In addition, this effect (high perceived cognitive trust in virtual context) is often amplified. Most companies do not even disclose to their customers that they are talking to a virtual actor. Although on the other hand, there exists a doubt on "affective trust" experienced by consumers in online service interactions. However, recently advanced AI tools like "Avatars," often equipped with empathetic intelligence, signal the prospect of "affective trust development" within virtual service encounters (Bente, Rüggenberg, Krämer, & Eschenburg, 2008).

- Another critical issue is the lack of social support because of stigma evolved during the pandemic. This is tackled up to a certain extent through AI-based interactive platforms that provide informational and social support to consumers (Coulson, 2005; Pfeil & Zaphiris, 2009). Even the healthcare communities act as a social resource for healthcare consumers where other patients and extended network share information and their own healthcare experiences (Nambisan, 2011).

- Lack of reciprocity is always a problem in healthcare due to information asymmetry. AI faces difficulty in solving this problem, as consumers often doubt the service interaction with Chatbots. However, if AI is planned effectively, for example, if a Chatbot discloses important information about itself, then there is a high probability that the consumer will also disclose his key facts with the machine and reciprocate positively (Chattaraman, Kwon, Gilbert, & Ross, 2019; Nambisan, 2011).

- Negative ethical perceptions imbibed in the mind of consumers are also a key concern within service interactions. Although AI still seems inexpert to resolve this issue. AI is itself encapsulated with ethical concerns like consumer's concern for data privacy and conflict of interest (Dignum,

2018). Focusing on this line, Fukawa and Erevelles (2014) explore the role of service provider's morals in service delivery's perceived reasonableness. However, establishing this moral or ethical norm within service provision is a challenging task (Ruane & Nallur, 2020).

Apart from AI playing a crucial role in combating uncertainty (because of the pandemic), it is also supporting VCC in healthcare services. This is reflected in Kaartemo and Helkkula (2018), which reviews the role of AI and robots in VCC. The researchers categorize the AI literature into four themes: generic field advancement, supporting service providers, enabling resource integration between service providers and beneficiaries, and supporting beneficiaries' wellbeing. Van Doorn et al. (2017) and Fan, Wu, Laurie, and Mattila (2016) explicitly mention that AI can assist the actors in integrating the resources during VCC. To better understand in context to healthcare, let us take an example: suppose some person meets with an accident and reaches the trauma center. In this case the patient's close one (care person) provides information about how the damage happened and his prior medical condition. This information (which is an essential resource for cocreation) is used for the early diagnosis and deciding the line of treatment. However, there is a high probability of this information being wrongly integrated. This is where AI can help. Using brain—computer interfaces, it is possible to decode patients' neural activities, and information can be obtained from the direct source, that is, patient (Bresnick, 2018). Moving on the same line, Čaić et al. (2018) recently explore the different ways AI (significantly socially assisted robots) can assist the vulnerable healthcare consumers (old age patients). Čaić et al. (2018) mainly identify three crucial ways, that is, providing security, social support, and cognitive support to old age patients. VCC literature within healthcare (Sweeney et al., 2015; Chen et al., 2020) asserts that knowledge and skills are the critical operant resources required for effective cocreation. AI directly affects both of these resources, like advanced mobile applications can enhance actor's healthcare knowledge (e.g., drug compliance knowledge) and skills (self-diagnosis skills) as evident in the literature (Barrett et al., 2019; Davey & Grönroos, 2019). The healthcare services interaction occurs at multiple levels, such as individual, dyadic, and an ecosystem (Sweeney et al., 2015). This essential prerequisite of cocreation is the interaction among involved actors, which is directly influenced by the level of AI used in healthcare services (Lee, 2019). AI-based technical devices or advanced applications can trigger the patient's interaction with essential healthcare services (Hoyer, Chandy, Dorotic, Krafft, & Singh, 2010; Lee, 2019). After the interaction, patients' engagement level directly affects their cocreation with the healthcare service providers (Hardyman, Daunt, & Kitchener, 2015). AI-based self-care tools (like wrist bands measuring calories, respirometer to keep daily records of asthma condition) provoke healthcare consumers to take part in their wellbeing actively and subsequently enhance the patient engagement (Davenport & Kalakota, 2019; Triberti & Barello, 2016). Finally, the patient experience directly affects their cocreation level and overall wellbeing outcome, as evident in the literature (Hardyman et al., 2015; Osei-Frimpong et al., 2015). In the same line, Daouk-Öyry, Alameddine, Hassan, Laham, and Soubra (2018) identify three essential pillars of patient experience, that is, employees, processes, and setting. All these components (employees, processes, and settings) are seen positively influenced by AI applications looking in AI literature. For example, the customized message (sent by AI-based Chatbots) received by the patient directly in his/her mobile phone enhances their medical care experience. The EHR shared across caregivers enhances the communication quality within service encounters, ultimately improving the patient perception of interpersonal relationships and perceived quality of care (Werder, 2015). All the above discussion shows how the varied forms of AI support healthcare VCC.

6.8   The spill-over effect of Artificial Intelligence

Based on the above discussions, it is clear that AI plays an augmented role in healthcare services, especially during a pandemic. However, there is another side of the picture as well. AI can result in negative outcomes and hinder the effective cocreation of value. This is reflected in recent studies that talk about AI and value codestruction (Čaić et al., 2018; Canhoto & Clear, 2020; Castillo, Canhoto, & Said, 2020; Neuhofer, 2016; Smith, 2013). Plé (2016) itself has elaborated on the reasons for value codestruction in a service setting, who is the pioneer of value codestruction (Plé & Chumpitaz Cáceres, 2010). Laud et al. (2019) explain the process of resource misintegration to elucidate the codestruction process in services. Laud et al. (2019) mention that customers may experience a resource loss or can unintentionally act as resource disintegrator. Interestingly, Castillo et al. (2020) observe a similar phenomenon of resource deficiency/resource loss and resource misintegration in AI-empowered service interactions. For example, the author noted that whenever the customer doubts the interaction (like if he doubts if he is talking to Chatbot or human), he may feel deceptive, resulting in deceptive integration of resources. Similarly, whenever the customer talks to Chatbot, he may feel unnecessary to be expressive in communication. The machine is smart and can automatically understand the issue, resulting in an unwillingness to integrate the resources (Castillo et al., 2020). Talking particularly in healthcare, Čaić et al. (2018) noted in their study that elderly patients often develop a feeling of "giving up" or "losing control" in their treatment because of the use of the socially assisted robot in their home. The author terms this as the "deactivator role of AI," that is, AI deactivates the focal actor (patient) by reducing their engagement. The same study also mentions another negative aspect of AI at an individual level, that is, "AI acting as an intruder." Here, intruder means a machine trying to invade the patient's privacy by continuous monitoring. This results in the skeptical response of elderly patients by contributing to a low level of personal resources for VCC (Čaić et al., 2018).

During the pandemic time, the issues with AI (as discussed above like consumer privacy issue, reducing consumer engagement, consumer loss of resources, actor's unwillingness to integrate the resources, deceptive integration of resources) could be amplified as reflected in AI studies (Hu et al., 2020; Naudé, 2020). The consumer often looks for more trust, ethics, personal identity, privacy, social support, human touch, and individual value protection in service interactions (Holt, 2020). However, AI may not seem to play a fair role in each of these aspects. Therefore, we will look into these aspects and try to find if AI can offer it:

- Trust: Miller (2019) argued that AI machines are mostly designed by researchers to interact in an ideal manner. However, the human being does not behave ideally; there is always some bias and social expectation attached to human behavior. This is the reason that AI fails to generate trust among consumers during service interactions (Miller, 2019).
- Ethics: Criticizing further recently, Carter et al. (2020) argue that AI, although claimed to be neutral, actually encodes values inevitably, which are generally very difficult to discern. One such example is ethical bias developed by AI machines by automatically learning the pattern from the dataset. It means if the particular section of a patient community like transgender cancer cases is underrepresented within AI datasets, then AI systems will automatically start producing biased outputs (less diagnosis of breast cancer cases for transgender patients) for the same patients (Carter et al., 2020).
- Personal identity: Buchanan-Oliver and Cruz (2011), while relating technology with liminality, claim that consumer's understanding of machine depends upon his understanding of himself and the increased use of technology (like technology as a prosthesis) can threaten the personal identity of "consumer as human" and ignite the fearful reactions in consumers.
- Privacy: Talking about the pitfalls of AI during COVID-19, Naudé (2020) recently argued that AI could hamper citizen's privacy as governments can continue this extraordinary surveillance of their citizens even after the pandemic.
- Social support: Emphasizing the importance of social support within healthcare services, researchers assert that AI can support medical services but cannot replace the social support offered by service providers, that is, doctors, nurses, and other paramedical staff (Vuorimies, Rosenius, Nirkkonen, Haakana, & Kuittinen, 2019).
- Individual actor's value: Vuorimies et al. (2019) claim that AI can hurt the healthcare service provider's value, that is, the value of autonomy.

Generally, doctors love to rely on their intuition and skills to diagnose the disease. If they feel like their freedom to use this skill is threatened due to AI, they may negatively react toward AI, which directly affects patient wellbeing.

- Human touch/support: It is recognized that although AI can resolve many issues during a pandemic (like offering contactless services through robots, faster medical diagnosis etc), but may also negatively affect the dynamics of human touch or empathy which is required for the mental wellbeing of recovering patients (Figueroa & Aguilera, 2020). Although the government has made different mental healthcare apps offering emotional support to patients, but its not effective because the vulnerable patients (old age patients) are not competent enough to use the smartphone apps (Figueroa & Aguilera, 2020).

6.9 Conclusion and future work

Overall this study made progress in highlighting the emerging role of AI (both positive and negative) in healthcare service interactions. While elaborating this service interaction, it focuses on the VCC process rooted in SDL. The study uses the context of the pandemic environment and pinpoints several uncertainties faced by healthcare services. The study elucidates the key uncertainties that could be managed by AI tools and result in the wellbeing of actors. In parallel, it also highlights the challenges that could not be solved effectively or entirely by the present AI tools. Although the study does not claim AI to be technically inferior, it only comments on its usage/perception/effect on the consumer using the social science approach. The study contributes knowledge to existing work by exploring the role of AI in cocreative healthcare interactions, especially during a pandemic. In addition, it highlights the spill-over effect of AI while consuming services. It mentions several ways AI-empowered interactions can fail to cope with the evolving uncertainty during COVID-19. However, the study focuses on COVID-19 but is equally applicable to any other pandemic or natural disaster. Studies also indicate a good number of ways; the present AI-based technology helps the healthcare service system cope with uncertainty during the pandemic. It can act as a guiding tool for policymakers and healthcare industrialists associated with planning or implementing AI usage (within healthcare) during a pandemic.

This study is restricted to the healthcare sector, but similar work could be explored in the education sector as both areas share a common trait, that is, credence nature of services. This study has used AI as a broader concept; future work could be focused on specific AI tools like robots or Chatbot. This work is purely conceptual. Future work could be planned to test the empirical relationships (like the relation between AI-based trust and cocreation intention, the relation between perceived uncertainty and AI-originated trust).

References

Arora, N. K. (2003). Interacting with cancer patients: The significance of physicians' communication behavior. *Social Science & Medicine*, *57*(5), 791−806. Available from https://doi.org/10.1016/s0277-9536(02)00449-5.

Barile, S., Saviano, M., & Polese, F. (2014). Information asymmetry and co-creation in health care services. *Australasian Marketing Journal (AMJ)*, *22*(3), 205−217. Available from https://doi.org/10.1016/j.ausmj.2014.08.008.

Barrett, M., Boyne, J., Brandts, J., Brunner-La Rocca, H. P., De Maesschalck, L., De Wit, K., & Hageman, A. (2019). Artificial intelligence supported patient self-care in chronic heart failure: A paradigm shift from reactive to predictive, preventive and personalised care. *EPMA Journal*, 1−20. Available from https://doi.org/10.1007/s13167-019-00188-9.

Bente, G., Rüggenberg, S., Krämer, N. C., & Eschenburg, F. (2008). Avatar-mediated networking: Increasing social presence and interpersonal trust in net-based collaborations. *Human Communication Research*, *34*(2), 287−318. Available from https://doi.org/10.1111/j.1468-2958.2008.00322.x.

Berry, L. L., & Bendapudi, N. (2007). Health care: A fertile field for service research. *Journal of Service Research*, *10*(2), 111−122. Available from https://doi.org/10.1177/1094670507306682.

Bhatnagar, V., Poonia, R. C., Nagar, P., Kumar, S., Singh, V., Raja, L., & Dass, P. (2020). Descriptive analysis of COVID-19 patients in the context of India. *Journal of Interdisciplinary Mathematics*, 1−16. Available from https://doi.org/10.1080/09720502.2020.1761635.

Bloom, G., Standing, H., & Lloyd, R. (2008). Markets, information asymmetry, and health care: Towards new social contracts. *Social Science & Medicine*, *66*(10), 2076−2087. Available from https://doi.org/10.1016/j.socscimed.2008.01.034.

Brashers, D. E. (2001). Communication and uncertainty management. *Journal of Communication*, *51*(3), 477−497. Available from https://doi.org/10.1111/j.1460-2466.2001.tb02892.x.

Bresnick, J. (2018). Top 12 ways artificial intelligence will impact healthcare. *Tools & strategies news, Health IT Analytics xtelligent Healthcare media* (Accessed online). https://healthitanalytics.com/news/top-12-ways-artificial-intelligence-will-impact-healthcare.

Buchanan-Oliver, M., & Cruz, A. (2011). Discourses of technology consumption: ambivalence, fear, and liminality. In T. L. C. R. Ahluwalia, & R. K. R. Duluth (Eds.), *NA − Advances in Consumer Research* (Vol. 39, pp. 287−291). Association for Consumer Research, Accessed online. Available from https://www.acrwebsite.org/volumes/1009170.

Butler, K. (1991). Toward understanding and measuring conditions of trust: Evolution of a conditions of trust inventory. *Journal of Management*, *17*, 643−663. Available from https://doi.org/10.1177/014920639101700307.

Čaić, M., Odekerken-Schröder, G., & Mahr, D. (2018). Service robots: Value co-creation and co-destruction in elderly care networks. *Journal of Service Management*, *29*(2), 178−205. Available from https://doi.org/10.1108/JOSM-07-2017-0179.

Canhoto, A. I., & Clear, F. (2020). Artificial intelligence and machine learning as business tools: A framework for diagnosing value destruction potential. *Business Horizons*, *63*(2), 183−193. Available from https://doi.org/10.1016/j.bushor.2019.11.003.

Carter, S. M., Rogers, W., Win, K. T., Frazer, H., Richards, B., & Houssami, N. (2020). The ethical, legal and social implications of using artificial intelligence systems in breast cancer care. *The Breast*, *49*, 25−32. Available from https://doi.org/10.1016/j.breast.2019.10.001.

Castillo, D., Canhoto, A. I., & Said, E. (2020). The dark side of AI-powered service interactions: Exploring the process of co-destruction from the customer perspective. *The Service Industries Journal*, 1−26. Available from https://doi.org/10.1080/02642069.2020.1787993.

Chattaraman, V., Kwon, W. S., Gilbert, J. E., & Ross, K. (2019). Should AI-based, conversational digital assistants employ social- or task-oriented interaction style? A task-competency and reciprocity perspective for older adults. *Computers in Human Behavior*, *90*, 315−330. Available from https://doi.org/10.1016/j.chb.2018.08.048.

Chen, T., Dodds, S., Finsterwalder, J., Witell, L., Cheung, L., Falter, M., & McColl-Kennedy, J. R. (2020). Dynamics of wellbeing co-creation: A psychological ownership perspective. *Journal of Service Management*. Available from https://doi.org/10.1108/JOSM-09-2019-0297.

Combi, C. (2017). Editorial from the new editor-in-chief: Artificial intelligence in medicine and the forthcoming challenges. *Artificial Intelligence in Medicine*, *76*, 37. Available from https://doi.org/10.1016/j.artmed.2017.01.003.

Coulson, N. S. (2005). Receiving social support online: An analysis of a computer-mediated support group for individuals living with irritable bowel syndrome. *Cyberpsychology & Behavior*, *8*(6), 580−584. Available from https://doi.org/10.1089/cpb.2005.8.580.

Daouk-Öyry, L., Alameddine, M., Hassan, N., Laham, L., & Soubra, M. (2018). The catalytic role of Mystery Patient tools in shaping patient experience: A method to facilitate value cocreation using action research. *PLoS One*, *13*(10), e0205262. Available from https://doi.org/10.1371/journal.pone.0205262.

Darby, M., & Karni, E. (1973). Free competition and the optimal amount of fraud. *Journal of Law and Economics*, *16*(1), 67−88. Accessed online: from http://www.jstor.org/stable/724826.

Davenport, T., & Kalakota, R. (2019). The potential for artificial intelligence in healthcare. *Future Healthcare Journal*, *6*(2), 94−98. Available from https://doi.org/10.7861/futurehosp.6-2-94.

Davey, J., & Grönroos, C. (2019). Health service literacy: Complementary actor roles for transformative value co-creation. *Journal of Services Marketing*, *33*(6), 687−701. Available from https://doi.org/10.1108/JSM-09-2018-0272.

D'Cruz, M. J., & Kini, R. B. (2007). *The effect of information asymmetry on consumer driven health plans*, . *Integration and innovation orient to e-society* (Volume 1, pp. 353−362). Boston, MA: Springer Accessed online. Available from https://doi.org/10.1007/978-0-387-75466-6_40.

Dignum, V. (2018). Ethics in artificial intelligence: Introduction to the special issue. *Ethics and Information Technology*, *20*(1), 1−3. Available from https://doi.org/10.1007/s10676-018-9450-z.

Duplaga, M., Zieliński, K., & Ingram, D. (Eds.), (2004). *Transformation of healthcare with information technologies* (Vol. 105). IOS Press. Available from http://doi.org/10.17148/IARJSET.2017.4520.

Dwivedi, Y. K., Hughes, L., Ismagilova, E., Aarts, G., Coombs, C., Crick, T., & Galanos, V. (2019). Artificial Intelligence (AI): Multidisciplinary perspectives on emerging challenges, opportunities, and agenda for research, practice and policy. *International Journal of Information Management*, 101994. Available from https://doi.org/10.1016/j.ijinfomgt.2019.08.002.

Esaiasson, P., Sohlberg, J., Ghersetti, M., & Johansson, B. 2020. How the coronavirus crisis affects citizen trust in government institutions and in unknown others—Evidence from "the Swedish Experiment." https://doi.org/10.31235/osf.io/6yw9r

Fan, A., Wu, L., (Laurie)., & Mattila, A. S. (2016). Does anthropomorphism influence customers' switching intentions in the self-service technology failure context? *Journal of Services Marketing*, *30*(7), 713−723. Available from https://doi.org/10.1108/JSM-07-2015-0225.

Figueroa, C. A., & Aguilera, A. (2020). The need for a mental health technology revolution in the COVID-19 pandemic. *Frontiers in Psychiatry*, *11*, 523. Available from https://doi.org/10.3389/fpsyt.2020.00523.

Finsterwalder, J., & Kuppelwieser, V. G. (2020). Equilibrating resources and challenges during crises: A framework for service ecosystem wellbeing. *Journal of Service Management*. Available from https://doi.org/10.1108/JOSM-06-2020-0201, ahead-of-print.

Frow, P., McColl-Kennedy, J. R., & Payne, A. (2016). Co-creation practices: Their role in shaping a health care ecosystem. *Industrial Marketing Management*, *56*, 24–39. Available from https://doi.org/10.1016/j.indmarman.2016.03.007.

Fukawa, N., & Erevelles, S. (2014). Perceived reasonableness and morals in service encounters. *Journal of Business Ethics*, *125*(3), 381–400. Available from https://doi.org/10.1007/s10551-013-1918-5.

Gopalan, H. S., & Misra, A. (2020). COVID-19 pandemic and challenges for socio-economic issues, healthcare and national health programs in India. *Diabetes and Metabolic Syndrome: Clinical Research and Reviews*, *14*(5), 757–759. Available from https://doi.org/10.1016/j.dsx.2020.05.041.

Han, P. K., Klein, W. M., & Arora, N. K. (2011). Varieties of uncertainty in health care: A conceptual taxonomy. *Medical Decision Making*, *31*(6), 828–838. Available from https://doi.org/10.1177/0272989X11393976.

Hardyman, W., Daunt, K. L., & Kitchener, M. (2015). Value co-creation through patient engagement in health care: A micro-level approach and research agenda. *Public Management Review*, *17*(1), 90–107. Available from https://doi.org/10.1080/14719037.2014.881539.

Hau, L. N. (2019). The role of customer operant resources in health care value creation. *Service Business*, *13*, 457–478. Available from https://doi.org/10.1007/s11628-018-00391-0.

Holt, G. R. (2020). The pandemic effect: Raising the bar for ethics, empathy, and professional collegiality. *Otolaryngology—Head and Neck Surgery*. Available from https://doi.org/10.1177/0194599820933179, 0194599820933179.

Hoyer, W., Chandy, R., Dorotic, M., Krafft, M., & Singh, S. (2010). Consumer co-creation in new product development. *Journal of Service Research*, *13*(3), 283–296. Available from https://doi.org/10.1177/1094670510375604.

Hu, Y., Jacob, J., Parker, G. J., Hawkes, D. J., Hurst, J. R., & Stoyanov, D. (2020). The challenges of deploying artificial intelligence models in a rapidly evolving pandemic. *Nature Machine Intelligence*, *2*, 298–300. Available from https://doi.org/10.1038/s42256-020-0185-2.

Huang, M. H., & Rust, R. T. (2018). Artificial intelligence in service. *Journal of Service Research*, *21*(2), 155–172. Available from https://doi.org/10.1177/1094670517752459.

Ives, J., Greenfield, S., Parry, J. M., Draper, H., Gratus, C., Petts, J. I., & Wilson, S. (2009). Healthcare workers' attitudes to working during pandemic influenza: A qualitative study. *BMC Public Health*, *9*(1), 1–13. Available from https://doi.org/10.1186/1471-2458-8-192.

Janamian, T., Crossland, L., & Wells, L. (2016). On the road to value co-creation in health care: The role of consumers in defining the destination, planning the journey and sharing the drive. *Medical Journal of Australia*, *204*(S7), S12–S14. Available from https://doi.org/10.5694/mja16.00123.

Kaartemo, V., & Helkkula, A. (2018). A systematic review of artificial intelligence and robots in value co-creation: Current status and future research avenues. *Journal of Creating Value*, *4*(2), 211–228. Available from https://doi.org/10.1177/2394964318805625.

Kahn, C. E., Jr (2017). From images to actions: Opportunities for artificial intelligence in radiology. *Radiology*, *285*(3), 719–720. Available from https://doi.org/10.1148/radiol.2017171734.

Kanawattanachai, P., & Yoo, Y. (2002). The dynamic nature of trust in virtual teams. *Journal of Strategic Information Systems*, *11*(3), 187–213. Available from https://doi.org/10.1016/S0963-8687(02)00019-7.

Kaplan, A., & Haenlein, M. (2019). Siri, Siri, in my hand: Who's the fairest in the land? On the interpretations, illustrations, and implications of artificial intelligence. *Business Horizons, 62* (1), 15−25. Available from https://doi.org/10.1016/j.bushor.2018.08.004.

Karlsen, A., & Kruke, B. I. (2018). The role of uncertainty during the Ebola pandemic in Western Africa (2014−2016). *Journal of Extreme Events, 5*(01), 1850009. Available from https://doi.org/10.1142/S2345737618500094.

Khanna, S., Sattar, A., & Hansen, D. (2013). Artificial intelligence in health−The three big challenges. *The Australasian Medical Journal, 6*(5), 315. Available from https://doi.org/ 10.21767/AMJ.2013.1758.

Kim, J. (2019). Customers' value co-creation with healthcare service network partners. *Journal of Service Theory and Practice, 29*(3), 309−3287. Available from https://doi.org/10.1108/ JSTP-08-2018-0178.

King, E. B., Tonidandel, S., Cortina, J. M., & Fink, A. A. (2015). Building understanding of the data science revolution and I-O psychology. In S. Tonidandel, E. B. King, & J. M. Cortina (Eds.), *Big data at work: The data science revolution and organizational psychology* (pp. 1−15). New York: Routledge, Accessed online. Available from https://www.taylorfran-cis.com/books/e/9781315780504/chapters/10.4324/9781315780504-6.

Kotler, P. (1994). Reconceptualizing marketing: An interview with Philip Kotler. *European Management Journal, 12*(4), 353−361. Available from https://doi.org/10.2478/v10147-010-0031-3.

Kumari, R., Kumar, S., Poonia, R. C., Singh, V., Raja, L., Bhatnagar, V., & Agarwal, P. (2020). Analysis and predictions of spread, recovery, and death caused by COVID-19 in India. *Big Data Mining and Analytics*. Available from https://doi.org/10.26599/BDMA.2020.9020013.

Kuppelwieser, V. G., & Finsterwalder, J. (2016). Transformative service research and service dominant logic: Quo Vaditis? *Journal of Retailing and Consumer Services, 28*, 91−98. Available from https://doi.org/10.1016/j.jretconser.2015.08.011.

Lalmuanawma, S., Hussain, J., & Chhakchhuak, L. (2020). Applications of machine learning and artificial intelligence for Covid-19 (SARS-CoV-2) pandemic: A review. *Chaos, Solitons & Fractals*. Available from https://doi.org/10.1016/j.chaos.2020.110059.

LaRosa, E., & Danks, D. (2018, December). Impacts on trust of healthcare AI. In *Proceedings of the 2018 AAAI/ACM conference on AI, ethics, and society* (pp. 210−215). https://doi.org/ 10.1145/3278721.3278771

Laud, G., Bove, L., Ranaweera, C., Leo, W. W. C., Sweeney, J., & Smith, S. (2019). Value co-destruction: A typology of resource misintegration manifestations. *Journal of Services Marketing, 33*(7), 866−889. Available from https://doi.org/10.1108/JSM-01-2019-0022.

Lee, D. (2019). Effects of key value co-creation elements in the healthcare system: Focusing on technology applications. *Service Business, 13*(2), 389−417. Available from https://doi.org/ 10.1007/s11628-018-00388-9.

Lee, R. G., Ozanne, J. L., & Hill, R. P. (1999). Improving service encounters through resource sensitivity: The case of health care delivery in an Appalachian community. *Journal of Public Policy & Marketing, 18*(2), 230−248. Available from https://doi.org/10.1177/ 074391569901800209.

Liang, H., Laosethakul, K., Lloyd, S. J., & Xue, Y. (2005). Information systems and health care—I: Trust, uncertainty, and online prescription filling. *Communications of the Association for Information Systems, 15*(1), 2. Available from https://doi.org/10.17705/ 1CAIS.01502.

Lusch, R. F., Vargo, S. L., & O'brien, M. (2007). Competing through service: Insights from service-dominant logic. *Journal of Retailing, 83*(1), 5−18. Available from https://doi.org/ 10.1016/j.jretai.2006.10.002.

MacPhail, T. (2010). A predictable unpredictability. The 2009 H1N1 pandemic and the concept of "strategic uncertainty" within global public health. *BEHEMOTH—A Journal on Civilisation, 3*(3), 57–77. Available from https://doi.org/10.1524/behe.2010.0020.

Mahomed, S. (2020). COVID-19: The role of artificial intelligence in empowering the healthcare sector and enhancing social distancing measures during a pandemic. *South African Medical Journal, 110*(7). Available from https://doi.org/10.7196/SAMJ.2020.v110i7.14841.

McColl-Kennedy, J. R., Vargo, S. L., Dagger, T. S., Sweeney, J. C., & Kasteren, Y. V. (2012). Health care customer value co-creation practice styles. *Journal of Service Research, 15*(4), 370–389. Available from https://doi.org/10.1177/1094670512442806.

Miller, T. (2019). Explanation in artificial intelligence: Insights from the social sciences. *Artificial Intelligence, 267*, 1–38. Available from https://doi.org/10.1016/j.artint.2018.07.007.

Nambisan, P. (2011). Information seeking and social support in online health communities: Impact on patients' perceived empathy. *Journal of the American Medical Informatics Association, 18*(3), 298–304. Available from https://doi.org/10.1136/amiajnl-2010-000058.

Naudé, W. (2020). Artificial intelligence vs COVID-19: Limitations, constraints and pitfalls. *AI & Society, 1*. Available from https://doi.org/10.1007/s00146-020-00978-0.

Neuhofer, B. (2016). *Value co-creation and co-destruction in connected tourist experiences. Information and communication technologies in tourism 2016* (pp. 779–792). Cham: Springer. Available from https://doi.org/10.1007/978-3-319-28231-2_56.

Nguyen, T. T. (2020). Artificial intelligence in the battle against coronavirus (COVID-19): A survey and future research directions. *Preprint*. Available from https://doi.org/10.13140/RG.2.2.36491.23846/1.

Osei-Frimpong, K., Wilson, A., & Owusu-Frimpong, N. (2015). Service experiences and dyadic value co-creation in healthcare service delivery: A CIT approach. *Journal of Service Theory and Practice, 25*(4), 443–462. Available from https://doi.org/10.1108/JSTP-03-2014-0062.

Ostrom, A. L., Fotheringham, D., & Bitner, M. J. (2019). *Customer acceptance of AI in service encounters: Understanding antecedents and consequences, . Handbook of service science* (Volume II, pp. 77–103). Springer. Available from https://doi.org/10.1177/1094670517752459.

Pfeil, U., & Zaphiris, P. (2009). Investigating social network patterns within an empathic online community for older people. *Computers in Human Behavior, 25*(5), 1139–1155. Available from https://doi.org/10.1016/j.chb.2009.05.001.

Plé, L. (2016). Studying customers' resource integration by service employees in interactional value co-creation. *Journal of Services Marketing, 30*(2), 152–164. Available from https://doi.org/10.1108/JSM-02-2015-0065.

Plé, L., & Chumpitaz Cáceres, R. (2010). Not always co-creation: Introducing interactional co-destruction of value in service-dominant logic. *Journal of Services Marketing, 24*(6), 430–437. Available from https://doi.org/10.1108/08876041011072546.

Prahalad, C. K., & Ramaswamy, V. (2004). Co-creating unique value with customers. *Strategy & Leadership, 32*(3), 4–9. Available from https://doi.org/10.1108/10878570410699249.

Priyanka Harjule, A. K., Agarwal, B., & Poonia, R. C. (2020). *Mathematical modeling and analysis of COVID-19 spread in India. Personal and ubiquitous computing (PAUC)*. Springer.

Quintal, V. A., Lee, J. A., & Soutar, G. N. (2010). Risk, uncertainty and the theory of planned behavior: A tourism example. *Tourism Management, 31*(6), 797–805. Available from https://doi.org/10.1016/j.tourman.2009.08.006.

Reddy, S., Fox, J., & Purohit, M. P. (2019). Artificial intelligence-enabled healthcare delivery. *Journal of the Royal Society of Medicine, 112*(1), 22–28. Available from https://doi.org/10.1177/0141076818815510.

Robertson, N., Polonsky, M., & McQuilken, L. (2014). Are my symptoms serious Dr Google? A resource-based typology of value co-destruction in online self-diagnosis. *Australasian Marketing Journal (AMJ), 22*(3), 246–256.

Robinson, S., Orsingher, C., Alkire, L., De Keyser, A., Giebelhausen, M., Papamichail, K. N., & Temerak, M. S. (2020). Frontline encounters of the AI kind: An evolved service encounter framework. *Journal of Business Research, 116,* 366–376. Available from https://doi.org/10.1016/j.jbusres.2019.08.038.

Ruane, E., & Nallur, V. (2020). "EHLO WORLD"–Checking if your conversational AI knows right from wrong. arXiv preprint arXiv:2006.10437.

Rubin, G. J., & Wessely, S. (2020). The psychological effects of quarantining a city. *British Medical Journal (online),* 368. Available from https://doi.org/10.1136/bmj.m313.

Sharma, P., Leung, T. Y., Kingshott, R. P., Davcik, N. S., & Cardinali, S. (2020). Managing uncertainty during a global pandemic: An international business perspective. *Journal of Business Research, 116,* 188–192. Available from https://doi.org/10.1016/j.jbusres.2020.05.026.

Sharma, S., Conduit, J., & Hill, S. R. (2017). Hedonic and eudaimonic wellbeing outcomes from co-creation roles: A study of vulnerable customers. *Journal of Services Marketing, 31*(4/5), 397–411. Available from https://doi.org/10.1108/JSM-06-2016-0236.

Sheth, J. N., & Uslay, C. (2007). Implications of the revised definition of marketing: From exchange to value creation. *Journal of Public Policy and Marketing, 26*(2), 302–307. Available from https://doi.org/10.1509/jppm.26.2.302.

Shiu, E. M., Walsh, G., Hassan, L. M., & Shaw, D. (2011). Consumer uncertainty, revisited. *Psychology & Marketing, 28*(6), 584–607. Available from https://doi.org/10.1002/mar.20402.

Singh, V., Poonia, R. C., Kumar, S., Dass, P., Agarwal, P., Bhatnagar, V., & Raja, L. (2020). Prediction of COVID-19 coronavirus pandemic based on time series data using Support Vector Machine. *Journal of Discrete Mathematical Sciences & Cryptography.* Available from https://doi.org/10.1080/09720529.2020.1784525.

Smith, A. M. (2013). The value co-destruction process: A customer resource perspective. *European Journal of Marketing, 47*(11/12), 1889–1909. Available from https://doi.org/10.1108/EJM-08-2011-0420.

Sweeney, J. C., Danaher, T. S., & McColl-Kennedy, J. R. (2015). Customer effort in value co-creation activities: Improving quality of life and behavioral intentions of health care customers. *Journal of Service Research, 18*(3), 318–335. Available from https://doi.org/10.1177/1094670515572128.

Tari Kasnakoglu, B. (2016). Antecedents and consequences of co-creation in credence-based service contexts. *The Service Industries Journal, 36*(1–2), 1–20. Available from https://doi.org/10.1080/02642069.2016.1138472.

Thesmar, D., Sraer, D., Pinheiro, L., Dadson, N., Veliche, R., & Greenberg, P. (2019). Combining the power of artificial intelligence with the richness of healthcare claims data: Opportunities and challenges. *Pharmacoeconomics, 37*(6), 745–752. Available from https://doi.org/10.1007/s40273-019-00777-6.

Topp, S. M., & Chipukuma, J. M. (2016). A qualitative study of the role of workplace and interpersonal trust in shaping service quality and responsiveness in Zambian primary health centres. *Health Policy and Planning, 31*(2), 192–204. Available from https://doi.org/10.1093/heapol/czv041.

Triberti, S., & Barello, S. (2016). The quest for engaging AmI: Patient engagement and experience design tools to promote effective assisted living. *Journal of Biomedical Informatics, 63,* 150–156. Available from https://doi.org/10.1016/j.jbi.2016.08.010.

Usher, K., Durkin, J., & Bhullar, N. (2020). The COVID-19 pandemic and mental health impacts. *International Journal of Mental Health Nursing*, 29(3), 315. Available from https://doi.org/10.1111/inm.12726.

Vafea, M. T., Atalla, E., Georgakas, J., Shehadeh, F., Mylona, E. K., Kalligeros, M., & Mylonakis, E. (2020). Emerging technologies for use in the study, diagnosis, and treatment of patients with Covid-19. *Cellular and Molecular Bioengineering*, 13(4), 249–257. Available from https://doi.org/10.1007/s12195-020-00629-w, Accessed online.

Van Doorn, J., Mende, M., Noble, S. M., Hulland, J., Ostrom, A. L., Grewal, D., & Petersen, J. A. (2017). Domo Arigato Mr. Roboto: Emergence of automated social presence in organizational frontlines and customers' service experiences. *Journal of Service Research*, 20(1), 43–58. Available from https://doi.org/10.1177/1094670516679272.

Vargo, S. L., & Lusch, R. F. (2004). Evolving to a new dominant logic for marketing. *Journal of Marketing*, 68(1), 1–17. Available from https://doi.org/10.1509/jmkg.68.1.1.24036.

Vargo, S. L., & Lusch, R. F. (2016). Institutions and axioms: An extension and update of service-dominant logic. *Journal of the Academy of Marketing Science*, 44(1), 5–23. Available from https://doi.org/10.1007/s11747-015-0456-3.

Verleye, K., Jaakkola, E., Hodgkinson, I. R., Jun, G. T., Odekerken-Schröder, G., & Quist, J. (2017). What causes imbalance in complex service networks? Evidence from a public health service. *Journal of Service Management*, 28(1), 34–356. Available from https://doi.org/10.1108/JOSM-03-2016-0077.

Virlée, J. B., Hammedi, W., & van Riel, A. C. R. (2020). Healthcare service users as resource integrators: Investigating factors influencing the co-creation of value at individual, dyadic and systemic levels. *Journal of Service Theory and Practice*, 30(3), 277–306. Available from https://doi.org/10.1108/JSTP-07-2019-0154.

Vuorimies, W., Rosenius, A., Nirkkonen, M., Haakana, J., & Kuittinen, J. (2019). The potentials and barriers of artificial intelligence in healthcare. Consolidated assignments from Spring 2019, 136 (Accessed online). https://pdfs.semanticscholar.org/41bf/840b08dc894ad-f0a53e4c0685d3a0378ee73.pdf#page = 138.

Wagner-Egger, P., Bangerter, A., Gilles, I., Green, E., Rigaud, D., Krings, F., & Clémence, A. (2011). Lay perceptions of collectives at the outbreak of the H1N1 epidemic: Heroes, villains and victims. *Public Understanding of Science*, 20(4), 461–476. Available from https://doi.org/10.1177/0963662510393605.

Werder, M. (2015). Health information technology: A key ingredient of the patient experience. *Patient Experience Journal*, 2(1), 143–147. Available from https://doi.org/10.35680/2372-0247.1071.

Xu, Y., Shieh, C. H., van Esch, P., & Ling, I. L. (2020). AI customer service: Task complexity, problem-solving ability, and usage intention. *Australasian Marketing Journal (AMJ)*. Available from https://doi.org/10.1016/j.ausmj.2020.03.005.

Zacher, H., & Rudolph, C. W. (2020). Individual differences and changes in subjective wellbeing during the early stages of the COVID-19 pandemic. *American Psychologist*. Available from https://doi.org/10.1037/amp0000702, Advance online publication.

Zandi, D., Reis, A., Vayena, E., & Goodman, K. (2019). New ethical challenges of digital technologies, machine learning and artificial intelligence in public health: A call for papers. *Bulletin of the World Health Organization*, 97(1), 2. Available from https://doi.org/10.2471/BLT.18.227686.

Chapter 7

The COVID-19 outbreak: social media sentiment analysis of public reactions with a multidimensional perspective

Basant Agarwal[1], Vaishnavi Sharma[2], Priyanka Harjule[3], Vinita Tiwari[4] and Ashish Sharma[1]

[1]Department of Computer Science and Engineering, Indian Institute of Information Technology (IIIT Kota), Kota, India, [2]Department of Chemical Engineering, National Institute of Technology, Raipur, India, [3]Department of Mathematics, Indian Institute of Information Technology (IIIT Kota), Kota, India, [4]Department of Electronics and Communications, Indian Institute of Information Technology (IIIT Kota), Kota, India

7.1 Introduction

The coronavirus disease (2019-nCoV) pandemic is an unprecedented crisis that has affected almost every individual in view of poverty, sustainability, and development all over the globe. Therefore, it becomes even more important to analyze such an effect on the development of people's lives (Cruz & Ahmed, 2018). After the spread of the SARS-COV-2 epidemic out of China, evolution in the coronavirus disease (2019-nCoV) pandemic shows dramatic differences among countries across the globe(Bhatnagar et al., 2020; Singh et al., 2020; Sohrabi et al., 2020). The panic of the COVID-19 outbreak has traversed around the globe that significantly impacted the global economy and the lifestyle of people all over the world (Qiu, Chen, Shi, & 2020, 2020). However, most of the countries adopted aggressive containment strategies of "lockdown" to mitigate the spread of the highly infectious COVID-19 disease (Chawla, Mittal, Chawla, & Goyal, 2020). The increase in the number of cases during the lockdown period has also spread the disruption, worry, stress, fear, disgust, sadness, and most importantly loneliness among the public at large (Zaroncostas, 2020).

On March 25, India, the world's second-most populous country with 1.3 billion citizens, witnessed the exhaustive confinement experiment in its

Cyber-Physical Systems. DOI: https://doi.org/10.1016/B978-0-12-824557-6.00013-3
117

history, in an endeavor to combat the COVID-19 disease (Roy et al., 2020; Sharma & Agarwal, 2020). India faced several challenges to deal with the increasing number of infected cases, mainly due to maintaining social distancing among its densely populated residential areas (Covid Social distancing, 2020). With firm preventive control measures and curtailments put in place by the Government of India in the form of nationwide lockdown, the citizens are experiencing a wide range of psychological and emotional responses such as fear and anxiety (Bao, Sun, Meng, Shi, Lu). In the same order, the Indian government imposed a complete lockdown on March 25, 2020, till April 14, 2020 (i.e., lockdown 1.0), which was extended till May 3, 2020 (lockdown 2.0), and further extended this in terms of lockdown 3.0 till May 17, 2020 (Lockdown in India, 2020). During all these lockdowns, the sentiments and psychological state of the public changed significantly, primarily due to various government policies imposed from time to time (Adam, 2020).

In the era of 24 h availability of various social media platforms and also due to social distancing, sentiment analysis plays a major role to understand public opinion. The Obama administration used sentiment analysis to measure public opinion before the 2012 presidential election. Various studies have shown social media is an important platform for information extraction. Sentiment analysis, also known as opinion mining, refers to the techniques and processes that help organizations or policymakers to retrieve information about how their people are reacting to a particular policy or situation.

With the worldwide spread of the COVID-19 infection and as a result of the lockdown and "work from home", individual activity on social media platforms such as Facebook, Twitter, and YouTube began to increase. Sentiment analysis provides some insights about various important issues from the perspective of common people. Through sentiment analysis, the decision is based on a significant amount of data rather than simple intuitions. It helps to understand the opinion of people about the situation or policy. We use social media cues to understand the relationship between people's sentiments and the effectiveness of the countermeasures deployed by the government during different phases of the complete lockdown in India. In a situation of a pandemic in a dynamic and developing country like India, policymakers cannot respond to every issue. Sentiment analysis in social media can help them prioritize the most important challenges (Agarwal & Mittal, 2016). This study aims at providing certain help in quickly identifying the policy priorities by examining and analyzing the sentiments of its people via a popular social media platform (Harjule, Gurjar, Seth, & Thakur, 2020; Zaroncostas, 2020).

Social media has been a major mode of communication among people that creates a large amount of user-generated opinionated textual data (Garrett, 2020). This huge collection of the data is a rich source for understanding the variations in the sentiment of the public toward different policies in India (Bhat et al., 2020). We use one of the most popular social media platforms—Twitter—to gauge the feelings of Indians toward the lockdown and other government policies.

A large amount of opinionated text on social media can provide important insights into understanding the public sentiments. In this chapter, we present the temporal sentiment analysis of Twitter data to understand the effect of lockdown on public perception. The impact of government control policies through various stages of lockdown in India is analyzed, which provides interesting aspects of public emotions. In addition, the analysis done in this study may offer a forward-thinking experience about the pandemic from the focal point of social media. Does this study provide answers to certain questions, such as how is Twitter being utilized to flow fundamental data and updates? What is the subject of conversation among individuals in the hour of corona emergencies? Results of lockdown and how individuals are responding to it? It was observed that initially, the positive sentiments were dominating, which was slowly shifted toward the negative sentiments by the end of the third lockdown.

This book chapter is organized as follows: Section 2 describes the data collection part. In Section 3, the impact of COVID-19 in the whole world is studied by finding out the countries that are most affected by the corona crisis. A choropleth map has been given to depict the same. The top four countries are compared from the number of cases point of view by plotting per capita graphs. Along with this sentiment analysis is done for the top five countries by extracting the tweets from those countries.

Further, in Section 4, data analysis of COVID-19 in India is carried out by understanding its trend through sentiment analysis of the tweets collected from India. This is visualized by making word cloud, scatter plot, and bar graphs. To do an in-depth analysis of COVID-19, the analysis for a particular city is done. The relationship between the number of tweets and days is done by using graphs. The frequency of most used words is also shown graphically and by word cloud. The sentiment analysis for the tweets from particular cities is done by the help of different types of plots such as box plot, scatter plot, polarity, and density curve. In Section 5, some of the most trending hashtags such as #WorkFromHome and #MigrantLabour are analyzed by doing sentiment analysis, word cloud, and by finding the frequency of words graphically. Finally, the conclusion is given in Section 6.

7.2   Data collection

To understand the public opinion and reaction over social media, we gathered tweets identified with COVID-19 utilizing Twitter's stream API from March 1, 2020 to May 26, 2020. A verified application associates with an open stream involving an example of the tweets being posted on Twitter. A channel showing which tweets are to be returned is remembered for each solicitation. For our exploration, the tweets were separated utilizing a run-down of explicit hashtags that we recognized dependent on the web look for mainstream COVID-19-related hashtags. The hashtags used for the analysis are as follows: #Lockdown, #StaySafe, #corona, #StayHome, #covid, #WorkFromHome, #indiafightscorona, #MigrantWorkers.

7.3 Sentiment analysis of the tweets collected worldwide

As on the condition of COVID-19 throughout the world till May 24, 2020, the worst-hit country due to COVID-19 is the United States, followed by the United Kingdom and Italy. From the graph as shown in Fig. 7.1, we can notice that within few days India is also going to enter in the list of the worst-hit countries due to COVID-19. To understand the trend of COVID-19, we will be analyzing these countries. As the population of the country varies, the effect of coronavirus on a country can be perfectly determined by the per capita graph. The population of Italy is least among all the mentioned countries; we can infer from the graph that it was having the most number of cases within 1 lakh people, and in that sense, COVID-19 impacted Italy the most.

Sentiment analysis helps us to know the observational thinking patterns of the people regarding a particular event. For this analysis, we collected the tweets with several hashtags related to the COVID-19 and coronavirus. All those tweets that mention COVID-19 or coronavirus in their text are collected. from these data we tried to know their sentiment, whether the tweets having COVID-19 as a keyword are positive, negative, or neutral. A glimpse of sample tweets has been shown in Table 7.1, which indicates how people have shown their views on social media. From Table 7.1, it should be noted

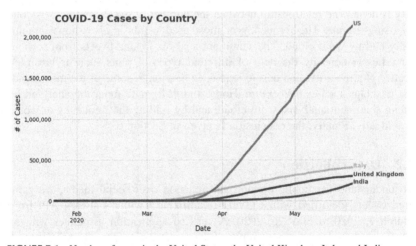

FIGURE 7.1 Number of cases in the United States, the United Kingdom, Italy, and India.

TABLE 7.1 Tweets collected from worldwide.

	Tweets	Polarity	Sentiment
1	Why are you hiding the number of people tested for COVID? You've been doing this for days now. Every time a statistic appears that shows your abject failure, it gets omitted. Are we living in North Korea?	−0.3166	Negative
2	Lovely lockdown treat @needaphone	0.5000	Positive
3	That's the entire point of using excess deaths as the metric. It means the method of counting COVID deaths doesn't influence the figures.	0	Neutral

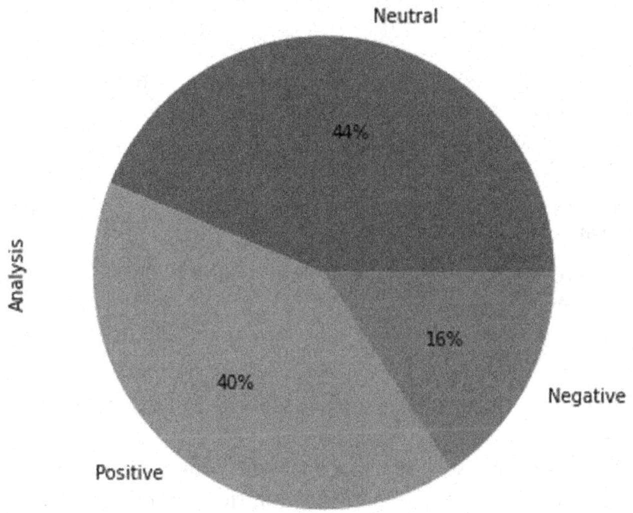

FIGURE 7.2 Percentage distribution of tweets worldwide.

that the polarity for tweet no. 1 is negative. This is because it consists of the negative word "failure" in this case. The sentiment of tweet no. 2 is positive as it involves the use of positive words. The word "lovely" is responsible for making the sentiment of tweet positive. The sentiment for the tweet no. 3 is neutral because it does not contain any specific word to make this tweet either positive or negative. Therefore it has neutral sentiment.

The pie chart shown in Fig. 7.2 denotes the percentage of tweets collected worldwide with positive, neutral, and negative sentiments. We can infer from the graph that most of the tweets are with neutral sentiments (around 44%), which are followed by positive sentiments (around 40%), and negative tweets are merely 16%. From the observation, it is concluded that

TABLE 7.2 Tweets from the United States.

	Tweets	Polarity	Sentiment
1	That was before he gave himself coronavirus.	0.000	Neutral
2	Cuomo admits "we all failed" at making coronavirus projections #FoxNews	−0.334	Negative
3	Today's coronavirus prayer, concerning our love of God	0.463	Positive

TABLE 7.3 Tweets from the United Kingdom.

	Tweets	Polarity	Sentiment
1	Awesome work being done by @CaringCity. Today they are generously delivering their #SoapAid to Edinburgh again to @SpaceBroomhouse. #NeverMoreNeeded.	0.5784	Positive
2	Agree re the scrums. Wrong in many ways—even before this coronavirus pandemic.	−0.235	Negative
3	Coronavirus is a once in a lifetime chance to reshape how we travel	0.000	Neutral

in this tough time, negativity is overshadowed by the positive attitude of people.

Next, sentiment analysis for the different countries is done separately. For this analysis, we had chosen the countries with the most number of cases. Table 7.2 depicts the sample tweets collected from the United States. Tweet no. 1 is of neutral sentiment, tweet no. 2 is with negative sentiment because it has the word "fail," which is responsible for making it negative. The tweet no. 3 is having the positive word "love" due to which it has positive sentiments. On similar grounds, Table 7.3 shows the sample tweets from the United Kingdom. Table 7.4 depicts the sample tweets from Italy. And, Table 7.5 consists of sample tweets from India.

Fig. 7.3 shows the graphs for sentiment analysis of the aforementioned countries. It can be inferred from the graphs (Fig. 7.3) that the country with the maximum number of tweets with positive sentiments is India, which is followed by Italy and the United States. Apart from this, the number of negative sentiments is almost the same in every country except the United States. The number of neutral sentiments is marginally high in the case of the United States. It can be observed from the graphs that the tweets in the

TABLE 7.4 Tweets from Italy.

	Tweets	Polarity	Sentiment
1	Enhancements will improve CPAC experience. #Cairns #Coronavirus #Council #Covid19 #Covidagenparl #Iorestoacasa	0.223	Positive
2	Are hand dryers effective in killing the ancient red pestilence? Women are also getting infected and spreading coronavirus to men.	−0.435	Negative
3	Coronavirus, la scoperta in Cina: tracce del virus	0.000	Neutral

TABLE 7.5 Tweets from India.

	Tweets	Polarity	Sentiment
1	India: Coping with the challenge of biomedical waste amidst coronavirus pandemic	0.00000	Neutral
2	Coronavirus recovery rate in India rises to over 42%, nearly 65,000 cured	0.1875543	Positive
3	Patience is extremely scarce in our country; that's why I fear #coronavirus will keep troubling us for plenty of months ahead.	−0.303456	Negative

United States vary from neutral to positive sentiments. In the case of the United Kingdom, the sentiment varies from neutral to negative. It clearly shows that either people are unable to cope up with COVID-19 effectively, or they are unhappy with the policies taken by the government for COVID-19 eradication, whereas, in India and Italy people have mixed attitudes toward coronavirus crises.

7.4 Sentiment analysis of Tweets for India

Sentiment analysis is contextual mining of text, which identifies and extracts subjective information in the source material and helps to understand the social sentiment of people while monitoring online conversations. It analyzes the given tweet and detects whether the underlying sentiment is positive, negative, or neutral. The tweets posted in India are crawled, and their corresponding subjectivity and polarity were computed. Polarity value lies in the range of $[-1,1]$ where 1 means positive statement and -1 means a negative statement. Subjective sentences generally refer to personal opinion, emotion, or judgment. Subjectivity is also a float that lies in the range of $[0,1]$. The

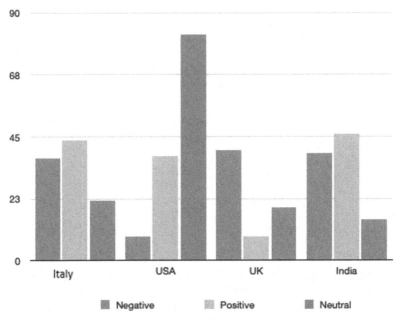

FIGURE 7.3 Sentiment analysis of different countries.

FIGURE 7.4 Word cloud of tweets collected in India.

sentiment analysis in India demonstrates that most of the public is positive even in a tough time due to COVID-19. Fig. 7.4 is the word cloud of all the tweets collected during this time in India, which shows the most used words by the people of India in the times of corona crises.

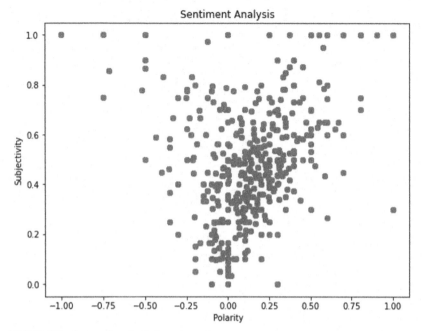

FIGURE 7.5   Scatter plot of polarity values of tweets.

As is seen from the word cloud (Fig. 7.4) that the most used word is *Stayhome StaySafe*, which is followed by *COVID-19*. These days people are tweeting more about the COVID-19, which clearly shows that it is the topic that is discussed the most among people of India. These tweets display a message, which is—stay home and stay safe and is the only measure to prevent ourselves from COVID-19. Further, Fig. 7.5 depicts the scatter plot that shows that the majority of the tweets are positive, as many points are on the right side of the polarity value 0.00.

The percentage of tweets with positive, negative, and neutral sentiments are observed as follows. The percentage of tweets with positive sentiments was 42.5%, and tweets with negative sentiments were 18.4%, while the rest was neutral tweets. Based on the initial 900 tweets gathered, Fig. 7.6 portrays the number of tweets with positive, negative, and neutral sentiments. From this, it can be inferred that the majority of tweets posted by individuals of India are more declined toward the positive side. This shows that no matter what, Indians were clear that they should consistently look toward the better side seeking for better days to come.

We also show a few of the sample tweets from the data collected during COVID-19 in India in Table 7.6.

It can be observed from the sample tweets that the public are positive mainly due to new initiatives, taking lockdown time for learning.

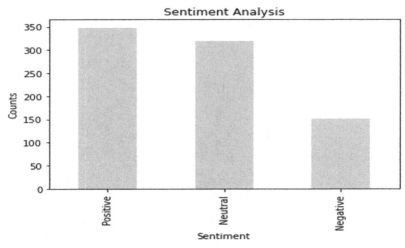

FIGURE 7.6 Sentiment analysis of tweets in India.

TABLE 7.6 Tweets collected from India.

	Tweets	Polarity	Sentiment	Subjectivity
1	That's a great initiative. Learning with the game. Amazing. #TogetherWeCan, #StayHomeStaySafe #StayHomeSaveLives #IndiaFightsCorona #worldfightscorona #StayHomeStayStrong #COVID19 #pandemic	0.476111	Positive	1.000
2	The trauma of #2014 & #2019 elections loss is still haunting you &your @INCIndia, which is very evident in your tweets & opinions. #COVID-19 is not the right time to show your frustration. #StayHomeStaySafe.	−0.26554	Negative	0.644
3	President Kenyatta hints at lifting the curfew and lockdown measures in the near future but, insists on individual responsibility to contain the spread of COVID-19.	0.000000	Neutral	0.000

7.4.1   COVID-19 analysis for individual city of India—Mumbai

To better understand the sentiment analysis in India, we further do more fine-grained analysis by considering only the most affected city of India, that is, Mumbai. The chart (Fig. 7.7) demonstrates the impact of COVID-19 on the individuals of Mumbai. The diagram is between the number of likes and the day on which the post is being liked. From the Fig. 7.7, it can be construed that most tweets identified with corona can be seen after March 15, when the COVID-19 came into the picture. It can be observed that the number of tweets related to corona saw an abrupt increment on the first day of lockdown, that is March 25 and on May 1 when the lockdown part 3 was declared by the prime minister of India. The scatter plot (Fig. 7.7) and line chart (Fig. 7.8) depicts the pattern by which tweets are posted. This trend helps us to predict which day most users will tweet.

The accompanying examination is on the number of tweets gathered from Mumbai. The investigation is done on the tweets relating to COVID-19 and its impact. The information is gathered from twitter from the hashtags, for example, *#Lockdown, #covid, and #corona*. Fig. 7.9 delineates the number of tweets gathered from March 20 to May 13. We can induce from the chart that the most talked about subject among the individuals of Mumbai is identified with the issue because of corona lockdown, which is accompanied by hashtags *#covid* and *#coronavirus*.

Moreover, a word cloud is drawn from the collected tweets and demonstrated in Fig. 7.10, in which the most used hashtags by the people of Mumbai from March 20 to May 13 are shown.

Fig. 7.11 shows the tweets relating to the utilization of various blends of words together, for example, how frequently the word lockdown and corona are used together. It likewise shows the occasions when a word is used alone.

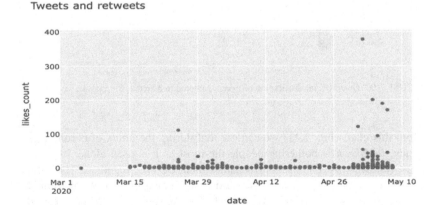

FIGURE 7.7   Scatter plot of the polarity of tweets in Mumbai.

Tweets and retweets

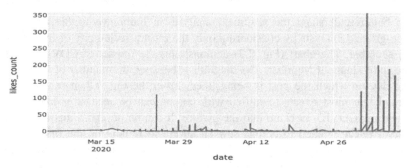

FIGURE 7.8 Line graph of the polarity of tweets in Mumbai.

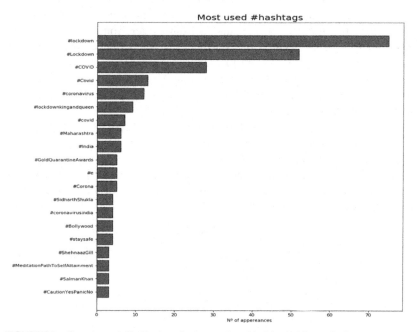

FIGURE 7.9 Quantity and distribution of tweets gathered in Mumbai for analysis.

From this chart (Fig. 7.11) we can see that during the hours of corona emergency, *lockdown* is bothering people the most as it is used alone 399 times, it is followed by the use of *lockdown* and *COVID* together. This graph depicts the trend of the most discussed topic among the people of Mumbai. Fig. 7.12 shows the words that are used most frequently by the people in the tweets.

FIGURE 7.10 The word cloud of the collected tweets.

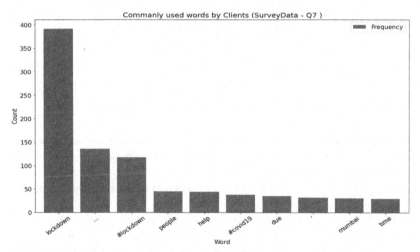

FIGURE 7.11 The occurrences of a frequent occurring hashtags.

7.4.1.1 Sentiment analysis of tweets in Mumbai

The sentiment analysis of the tweets of people of Mumbai is performed, and corresponding to subjectivity and polarity values are computed. Table 7.7 represents the sentiment, subjectivity, and polarity of a few sample representative tweets. The average polarity from all the tweets was collected, and that came out to be 0.080461, which shows that on average, most people are in between neutral to a positive opinion regarding the corona crises.

The covariance between the subjectivity and polarity is computed to be 0.058285491, its positive value suggests that variables change in the same

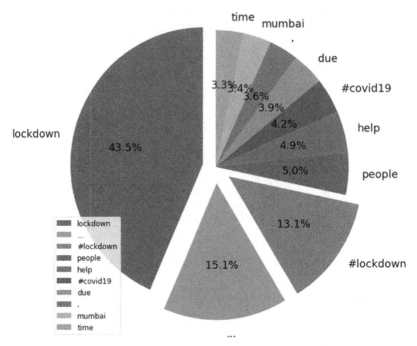

FIGURE 7.12 The pie-chart for the most used words among the tweets in Mumbai.

direction. In addition, the correlation is 0.74972, which shows that as subjectivity increases, the polarity in the response also increases. In other words, the more opinions/feelings are expressed, the more the overall positive sentiments will be. Fig. 7.13 shows the scatter diagram of the polarity and subjectivity of the collected tweets.

The boxplot of subjectivity and polarity along with polarity distribution of tweets are shown in Figs. 7.14 and 7.15. It can be observed that most of the tweets are with normal sentiments, around 120 times, which is followed by most positive sentiments. In addition, we show the density graph in Fig. 7.16, which shows that most of the tweets are dense at neutral polarity value.

7.5 Analysis of few most trending hashtags

So far, the analysis was based on geographical locations, it is also important to understand the sentiments of people concerning specific policies such as WorkFromHome, MigrantWorkers, etc. In this part, we will analyze some of the most trending hashtags such as #WorkFromHome, #MigrantWorkers, #MeTooMigrant, etc.

TABLE 7.7 Sample tweets with the sentiment, polarity, and subjectivity from Mumbai.

	Tweets	Polarity	Sentiment	Subjectivity
1	@indiahaier thank you very much for your assistance during such lockdown period we truly appreciate the same thanks to your team for giving solution	0.498655	Positive	0.6433333
2	almost 52000 cases are from Maharashtra Gujarat Delhi Tamilnadu these states have failed to tackle COVID and break the chain even after 55 days of lockdown	−0.165543	Negative	0.2888889
3	this lockdown @runcaralisarun	0.000000	Neutral	0.00000000

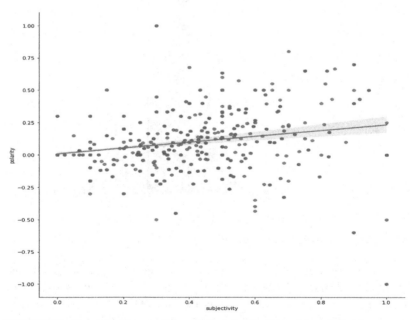

FIGURE 7.13 The scatter diagram of the polarity and subjectivity of the collected tweets.

7.5.1 Opinion analysis for the hashtag #WorkFromHome

The world is witnessing the impact of COVID-19 on all facets of life, in all countries, and all industries. No one is certain about how much and how

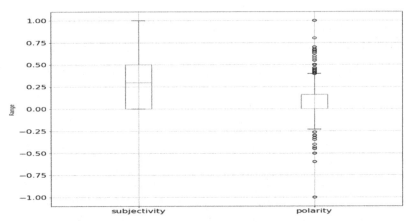

FIGURE 7.14   Boxplot of subjectivity and polarity of tweets in Mumbai.

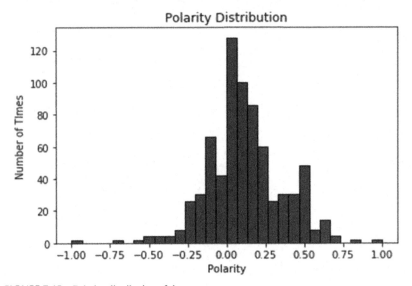

FIGURE 7.15   Polarity distribution of the tweets.

long the impact of the pandemic will last on the global economy. During the lockdown, the Indian IT industry made employees "Work from Home" as per the government's mandate. As a result, about 90% of employees worked from home with 65% of them from homes in metros and the rest 35% from homes in small towns.

To understand the attitude of people toward the work from home policy, we crawled the tweets in India. Few examples of tweets mentioning work from home are illustrated in Table 7.8, along with the sentiment and polarity.

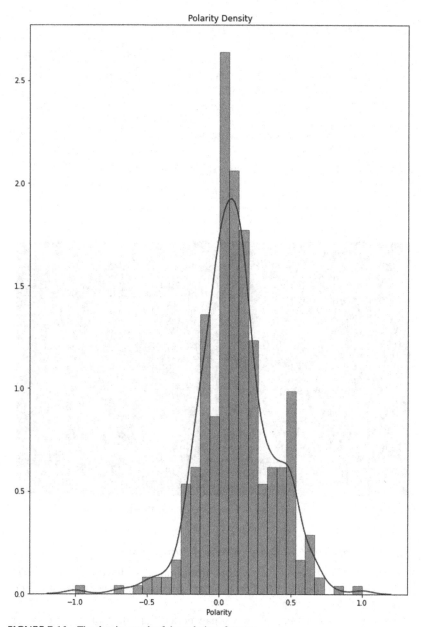

FIGURE 7.16 The density graph of the polarity of tweets.

The word cloud shown in Fig. 7.17 represents some of the most used words that were used in the tweets, along with hashtag work from home. From this word cloud, we can infer that the most used words are work and home.

TABLE 7.8 Sample tweets having hashtag #WorkFromHome.

	Tweets	Polarity	Sentiment
1	This is a lovely (funny but melancholy) piece about what we miss when we work from home	0.375000	Positive
2	Some of us do not want 2 risks getting sick & taking this home 2 family. If you go to a store, you can wear a mask for a 1/2 hr or get someone else to shop for you.	−0.357143	Negative
3	Office calls especially this work from home period	0.00000	Neutral

FIGURE 7.17 The word cloud of the tweets having hashtag #WorkFromHome.

7.5.1.1 Sentiment analysis of #WorkFromHome

For the analysis of the keyword work from home, we have gathered almost 1000 tweets that are classified in positive, negative, and neutral sentiments depicted in Fig. 7.18.

From the graph in Fig. 7.18, we can notice that most of the tweets are with the neutral sentiment (around 600), followed by positive sentiments (around 350), and least are the tweets with neutral sentiments (around 90). This shows that feedback on the work from home is good as most of the people are enjoying this new trend (Fig. 7.19). Fig. 7.20 represents the most frequent words used in tweets with these hashtags.

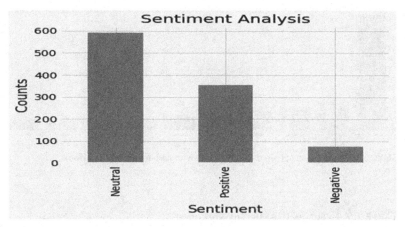

FIGURE 7.18   The sentiment analysis of the tweets having hashtag #WorkFromHome.

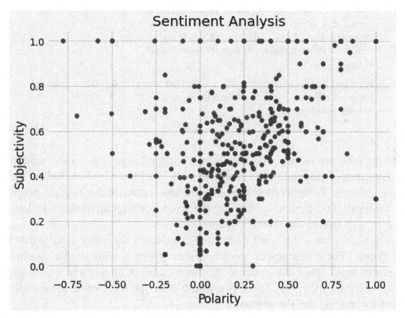

FIGURE 7.19   The scatter plot of the polarity values and subjectivity towards #WorkFromHome.

7.5.2   Sentiment analysis of #MigrantWorkers

India is dealing with another issue that is also challenging, such as dealing with coronavirus lockdown. India's nationwide lockdown amidst the COVID-19 pandemic has critically dislocated its migrant population.

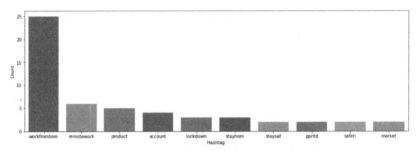

FIGURE 7.20   Most frequent words in the tweets with hashtag #WorkFromHome.

TABLE 7.9 Tweets with #MigrantWorkers.

	Tweets	Polarity	Sentiment
1	Excellent write up by @MukandRita "The way a nation treats their poor reveals the health, hearts, and mental state of the nation and if a nation is kind to their poor, that nation is rich in spirit, heart, and wealth." #MigrantWorkers #CoronaWarriorsIndia #MigrantCrisis	0.67500	Positive
2	Bombay HC Asks Maharashtra Govt To Submit Status Report of Migrants In Mumbai#BombayHC #Mumbai #MigrantWorkers	0.00000	Neutral

Lacking jobs and money, and with public transportation shut down, hundreds of thousands of migrants were forced to walk hundreds of miles back to their home villages. To understand the situation and sentiment for migrant people, we crawled 1000 tweets mentioning the hashtag #MigrantWorkers in India. Few of such tweets are shown in Table 7.9.

Fig. 7.21 shows the word cloud that consists of the most used words in the tweets. The frequency of word *migrantworkers* is highest. We can infer from the word cloud that most of the tweets consist of positive words such as wonderful, happy, etc. These tweets can either be more of an appreciation tweet for helping out the workers.

In the sentiment analysis of such tweets, it was observed that most of the tweets are with neutral sentiments, crossing 500, followed by positive tweets (just less than 500), and rest very few tweets are negative. It is observed that tweets were not showing negative sentiments because most of the people using Twitter social media platforms were a highly educated population, and very few migrant workers use it. And most of the tweets about the migrant workers are posted by them discussing what can be done for them, and also help out the workers. It can be intuitively thought that the negative sentiment

FIGURE 7.21 The word cloud of the tweets collected having hashtag #MigrantWorkers.

could not be seen over social media during this time, possibly because such people did not express their feelings on Twitter. The polarity of these tweets varies between 0 and above.

7.6 Conclusion

This chapter focuses on the analysis with the impact of COVID-19 based on the opinion expressed over social media. The COVID-19 has affected each country adversely, but few of them are suffering the most. The sentiment analysis was carried out in the chapter, which is a powerful tool to know the reaction of people toward an event. Most of the tweets gathered vary from neutral to positive sentiments starting from the pandemic with afterward. It means that people are moving forward with the optimistic approach in these tough times. The relation between the number of tweets and date was analyzed, which showed that most tweets were noted on the following day when important announcements were made, like that of lockdown. The chapter demonstrates the sentiment analysis concerning the geographical locations with time, in addition, to specific policy hashtags such as #WorkFromHome and #MigrantWorkers. The inferences drawn from these data help to get an insight into the mental space of the majority of people during the harsh conditions of lockdown in India. Further, it may also help the government to redesign and redefine the further course of actions in order to prevent the spread of COVID-19 while taking into account the valuable opinions of its people.

References

Adam, D. (2020). Special report: The simulations driving the world's response to COVID-19. *Nature, 580*(7803), 316−318.

Agarwal, B., & Mittal, N. (2016). *Prominent feature extraction for sentiment analysis. Springer Book Series: Socio-affective Computing series* (pp. 1−115). Springer International Publishing. Available from http://doi.org/10.1007/978-3-319-25343-5.

Bao, Y., Sun, Y., Meng, S., Shi, J., & Lu, L. (2020). 2019-nCoV epidemic: Address mental health care to empower society. *The Lancet, 395*, E37−E38. Available from https://doi.org/10.1016/S0140-6736(20)30309-3.

Bhat, M., Qadri, M., Beg, N. A., Kundroo, M., Ahanger, N., & Agarwal, B. (2020). Sentiment analysis of social media response on the Covid19 outbreak. *Brain, Behavior, and Immunity, 87*. Available from https://doi.org/10.1016/j.bbi.2020.05.006.

Bhatnagar, V., Poonia, R. C., Nagar, P., Kumar, S., Singh, V., Raja, L., & Dass, P. (2020). Descriptive analysis of COVID-19 patients in the context of India. *Journal of Interdisciplinary Mathematics*, 1−16.

Chawla, S., Mittal, M., Chawla, M., & Goyal, L. M. (2020). Corona virus − SARS-CoV-2: An insight to another way of natural disaster. *EAI Endorsed Transactions on Pervasive Health and Technology, 2020.*

Covid Social Distancing, https://www.bbc.com/news/world-asia-india-52393382, accessed on May 10, 2020.

Cruz, M., & Ahmed, S. A. (2018). On the impact of demographic change on economic growth and poverty. *World Development, 105*, 95−106.

Garrett, L. (2020). COVID-19: The medium is the message. *The Lancet, 395*(10228), 942−943. Available from https://doi.org/10.1016/S0140-6736(20)30600-0.

Harjule P., Gurjar A., Seth H., Thakur P. (2020). Text classification on Twitter data, 3rd International Conference on Emerging Technologies in Computer Engineering: Machine Learning and Internet of Things, pp. 160−164, Available from https://doi.org/10.1109/ICETCE48199.2020.9091774.

Lockdown in India. 2020. https://en.wikipedia.org/wiki/~COVID-19_pandemic_lockdown_in_India (Accessed on May 21, 2020).

Qiu, Y., Chen, X., & Shi, W. (2020). Impacts of social and economic factors on the transmission of coronavirus disease 2019 (COVID-19) in China. *Journal of Population Economics, 2020.* Available from https://doi.org/10.1007/S00148-020-00778-2.

Roy, D., Tripathy, S., Kar, S. K., Sharma, N., Verma, S. K., & Kaushal, V. (2020). Study of knowledge, attitude, anxiety & perceived mental healthcare need in Indian population during COVID-19 pandemic. *Asian Journal of Psychiatry, 51.*

Sharma, A., & Agarwal, B. (2020). A cyber-physical system approach for model based predictive control and modeling of COVID-19 in India. *Journal of Interdisciplinary Mathematics, 24.*

Singh, V., Poonia, R. C., Kumar, S., Dass, P., Agarwal, P., Bhatnagar, V., & Raja, L. (2020). Prediction of COVID-19 coronavirus pandemic based on time series data using support vector machine. *Journal of Discrete Mathematical Sciences & Cryptography, 23.*

Sohrabi, C., Alsafi, Z., O'Neill, N., Khan, M., Kerwan, A., Al-Jabir, A., ... Agha, R. (2020). World Health Organization declares global emergency: A review of the 2019 novel coronavirus (COVID-19). *International Journal of Surgery (London, England), 76*, 71−76. Available from https://doi.org/10.1016/j.ijsu.2020.02.034.

Zaroncostas, J. (2020). How to fight an infodemic. *The Lancet, 395*(10225). Available from https://doi.org/10.1016/S0140-6736(20)30461-X.

Chapter 8

A new approach to predict COVID-19 using artificial neural networks

Soham Guhathakurata[1], Sayak Saha[1], Souvik Kundu[2],
Arpita Chakraborty[3] and Jyoti Sekhar Banerjee[3]
[1]*Department of CSE, Bengal Institute of Technology, Kolkata, India, [2]Department of Electrical and Computer Engineering, Iowa State University, Ames, IA, United States, [3]Department of ECE, Bengal Institute of Technology, Kolkata, India*

8.1 Introduction

The coronavirus pandemic, also known as COVID-19 pandemic (Wikipedia, 2020) created a world-wide chaos since its advent in December 2019 in Wuhan, China. With over 1,06,94,288 confirmed cases across the world, the virus is spreading at an exponential rate. The World Health Organization stated that COVID-19 is by far the most severe global health emergency and declared it as a Public Health Emergency of International Concern on January 30, 2020 (Wikipedia, 2020). As of August 6, 2020, more than 188 countries and territories have reported more than 18.7 million cases of COVID-19. The transmission rate of the virus is significantly high and spreads mainly via respiratory droplets from coughing, sneezing, talking, and also close contacts. Like a typical airborne disease, the droplets remain suspended in the enclosed spaces, which also causes the transmission. The virus can even spread from a person who does not have any symptoms, which makes COVID-19 such a dangerous entity.

From the overall cases so far, the most common symptoms include fever, cough, and shortness of breath (Webmd, 2020). The conditions of the patient deteriorate in cases of pneumonia and acute respiratory distress syndrome (United States Centers for Disease Control and Prevention (CDC), 2020a). The onset of symptoms takes typically around five days from the time of exposure but may range from 2 to 14 days (United States Centers for Disease Control and Prevention (CDC), 2020b; Velavan and Meyer, 2020). Many medical organizations over the world are testing their vaccines to

Cyber-Physical Systems. DOI: https://doi.org/10.1016/B978-0-12-824557-6.00009-1
139

overcome this virus. So far, the standard method of diagnosis includes the real-time reverse transcription polymerase chain reaction, that is., rRT-PCR (Zu et al., 2020), which is performed on throat-swab specimens. To restrict the contamination of the coronavirus certain preventive measures are recommended, which include proper sanitization, wearing protective face masks in public, ensuring proper distance from other people, and maintaining self-isolation for people who suspect any hint of the coronavirus symptoms.

Till now, there is an absence of standard methods for predicting the risk level of a patient based on their early and mild symptoms. Since its advent, the virus has changed its character and evolved quite often, which has increased the unpredictability to judge the infection taking only a handful of symptoms. Therefore the authors in this chapter have attempted to predict COVID-19 tasking a list of symptoms and conditions based on their severity, frequency, and impact on the patients. The risk assessment has been categorized into three different classes, which are Not Infected, Mildly Infected, and Severely Infected. Keeping in mind the evolving nature of the virus, we have applied the Artificial Neural Network (ANN) to classify the risk into the mentioned classes. The use of ANN reduces the judgment of error while predicting the outcome, which again is an important feature taking into consideration the severe effect of coronavirus.

The chapter is constructed as follows: section 8.2 deals with related studies followed by the fundamental symptoms and conditions responsible for COVID-19 infection in section 8.3. Section 8.4 states the proposed COVID-19 detection methodology. In section 8.5, a brief description of ANNs is mentioned. Sections 8.6 and 8.7 display the prediction of COVID-19 infection risk using ANN and experimental results and discussion, respectively, followed by the performance comparison between ANN and other classification algorithms in section 8.8. Lastly, we present some concluding comments.

8.2 Related studies

ANNs have found applications in many disciplines that include system identification, pattern recognition, medical diagnosis, social network filtering, and much more. A brief review is presented here.

Jiang, Trundle, and Ren (2010) have used the ANN for resolving medical imaging problems with fixed structure and training procedures. They have analyzed, processed, and characterized medical images and resolved problems relevant to medical imaging by expanding neural networks.

Huang, Shen, and Duong (2010) developed a flexible ANN to predict ischemic tissue fate and permanent cerebral artery occlusion in rats. To improve prediction accuracy neighboring pixel information and infarction incidence were included in the ANN model. The major finding was the objective framework that the ANN predictive model can provide to evaluate stroke treatment on an individual patient basis.

Sinha and Wang (2008) developed the ANN prediction model to predict the values of maximum dry density, permeability, and optimum moisture content with classification properties of the soils. The model gave an accuracy of 95%.

Hsu, Gupta, and Sorooshian (1995) presented an effective alternative to the autoregressive moving average with exogenous inputs (ARMAX) time series approach using a new procedure for the ANN model for developing input − output simulation and forecasting models. This new model will be able to simulate the nonlinear hydrologic behavior of watersheds in situations that do not require modeling of the internal structure of the watershed.

Hill, Marquez, O'Connor, and Remus (1994) presented a thorough comparison between ANNs and statistical models. The topic of comparison was based on regression-based forecasting, time series forecasting, and decision making. The primary objective of this report is to provide an assessment of ANN for forecasting and decision-making models.

Lee, Cha, and Park (1992) applied an ANN to forecast the short-term load for a large power system. Various combinations of the neurons with one or two hidden layers were tested. The predicted results were then compared in terms of forecasting error. The conclusion of this chapter claims that a good load forecast is obtained when the neural networks are grouped into different load patterns.

Kumar et al. (2002) estimated the daily grass reference crop evapotranspiration (ETo) using ANNs. Based on the weighted standard error of estimate and minimal ANN architecture, the best ANN architecture was selected. The results were compared with the conventional methods (Penman − Monteith) and found to be much more satisfactory by using ANN to predict ETo.

William G. Baxt (Baxt, 1991) used an ANN to identify myocardial infarction in patients. The network was tested on 331 consecutive patients with anterior chest pain. The network was compared with that of physicians caring for the same patient. A better result was obtained by the ANN.

Guresen, Kayakutlu, and Daim (2011) evaluated the effectiveness of neural network models. The models that are analyzed include multilayer perceptron, hybrid neural network, and dynamic ANN. The models are then compared based on Mean Square Error and Mean Absolute Deviate. Kumari et al. (2020), Bhatnagar et al. (2020), and Singh et al. (2020) analyzed the current situation and forecasted some about future scenario.

8.3   Fundamental symptoms and conditions responsible for COVID-19 infection

During the incubation period of 14 days, some infected people show no symptoms while others show a wide range of symptoms. The two most common symptoms include fever (88%) and dry cough (68%). The less common symptoms include fatigue (38%), shortness of breath (19%), sputum production (33.4%), persistent chest pain, headache (14%), sore throat (13.9%), chills (11%), nasal congestion (4.8%), nausea (5%), diarrhea (4%), haemoptysis (0.9%), and pink eyes and lips (0.8%) (Huang et al., 2020; WHO − China Joint Mission, 2020; World Health Organization, 2020a). Though the recovery rate of the patients increased in the due course of time, the fatality rate of patients with cardiovascular disease and hypertension (Huang et al., 2020) has remained constant. The majority of the patients who suffered death due to COVID-19 progressed to the critical condition, which includes respiratory failure, septic shock, multiple organ failure (WHO − China Joint Mission, 2020; World Health Organization, 2020a; Chen et al., 2020; World Health Organization, 2020b). Patients with mild symptoms like common cold and cough tend to recover within 2 weeks (Zu et al., 2020). However, those patients suffering from severe symptoms like high respiratory frequency or dyspnea, in best cases recover within a span of 3 − 6 weeks.

The transmission rate of coronavirus is higher than that of influenza. Patients showing mild or nonspecific symptoms are highly infectious. Such patients remain infectious for an average of 2 weeks and can transmit the virus even from a single respiratory droplet. Studies have shown that people without symptoms also transmit the virus, which is known as asymptomatic transmission (U.S. Centers for Disease Control and Prevention CDC, 2020). The World Health Organization recommends 1 meter of social distance to restrict the primary spread of the virus from close contacts and from inhaling small droplets produced by an infected person by coughing or sneezing (Wang, Tang, & Wei, 2020; Rothan and Byrareddy, 2020; Zheng, Ma, Zhang, & Xie, 2020; Fang, Karakiulakis, & Roth, 2020; Guan et al., 2020; The Epidemiological Characteristics of an Outbreak of 2019 Novel Coronavirus Diseases COVID-19, 2020). The other modes of transmission include the suspension of the droplet in the air for quite a long period. The virus can also spread from the contaminated droplets that have fallen to floors or surfaces. Though the level of contamination that can transmit the infection via surfaces is not known, proper surface disinfectants should be used to restrict the unbounded spread of the virus.

8.4 Proposed COVID-19 detection methodology

Since the outbreak of COVID-19, the virus has evolved at a tremendous rate. One of the major causes of the failure of COVID-19 detection (Biswas, Sharma, Ranjan, & Banerjee, 2020; Guhathakurata, Kundu, Chakraborty, & Banerjee, 2020; Guhathakurata, Saha, Kundu, Chakraborty, & Banerjee, 2020a, 2020b) from the symptoms is mainly because of the uncertain nature of the virus. As a result of which no proper dataset is available to use as a reference. A synthetic multicriterion (Banerjee & Chakraborty, 2014, 2015;

Banerjee, Chakraborty, & Chattopadhyay, 2017, 2018b, 2018c, 2018a, 2021; Banerjee, Chakraborty, & Karmakar, 2013; Banerjee, Goswami, & Nandi, 2014; Banerjee & Karmakar, 2012; Saha, Chakraborty, & Banerjee, 2019; Saha et al., 2017) dataset has been coined by the authors to serve the purpose of this chapter, which is mentioned in the appendix (Table A1). The attributes or symptoms and conditions that have been taken into consideration include age, body temperature, dry cough, chest pain, breathing rate, hypertension, cardiovascular diseases, diabetes, tiredness, current smoker, and contact with a person with fever or cold in the last few days. The attributes are discussed below concerning their impact related to COVID-19:

1. Age: The fatality rate of the old aged people is more compared to the young ones in the coronavirus cases. The main reason for this development is basically due to the strong immunity level in the younger persons compared to the aged ones.
2. Body temperature: One of the most significant COVID-19 symptoms is body temperature. Temperature above 100°C is mild fever and above 103°C is high fever. Body temperature can be easily detected by the thermal scanner using sensors (Das, Pandey, Chakraborty, & Banerjee, 2017; Paul, Chakraborty, & Banerjee, 2019; Roy, Dutta, Biswas, & Banerjee, 2020; Paul et al., 2017; Das, Pandey, & Banerjee, 2016).
3. Dry cough: Cough without mucus or phlegm is one of the common symptoms of COVID-19.
4. Tiredness: The early sign of COVID-19 infection.
5. Chest pain: Persistent chest pain with breathing problems gives rise to a critical condition.
6. Nasal congestion: One of the symptoms to be looked upon as an early warning sign of coronavirus.
7. Runny nose: This is quite a common symptom of cold but when coupled with fever and breathing problems the person can be predicted into the early stages of coronavirus.
8. Sore throat: Considered to be the early symptom of coronavirus. Patients should consult the doctor when suffering from a sore throat.
9. Diarrhea: Patients suffering from diarrhea are very prone to be getting infected by a coronavirus.
10. Breathing rate: Breathing rate above 30 breath/min is a critical condition.
11. Hypertension: Patient with stage 1 hypertension (140−159 blood pressure) are less susceptible to coronavirus compared to a patient with stage 2 hypertension (160 and above blood pressure).
12. Cardiovascular diseases: The fatality rate increases for patients with cardiovascular diseases.
13. Diabetes: One of the critical symptoms to look upon to check for the risk of coronavirus infection.

14. Current smoker: Current smokers are more susceptible to suffer from acute respiratory disease syndrome.

15. Contact with a person with fever and cold in the last few days: This is a very important condition to be checked upon to track the source of coronavirus.

In this chapter, the authors have not considered some important factors like loss of smell and taste because nowadays it is almost the proven fact that loss of smell and taste is due to COVID-19. If anyone is having a loss of smell and taste, then he/she is advised to test for COVID-19. In this chapter, the authors have considered mainly symptomatic patients. As it is really hard to predict asymptomatic patients, a bulk number of people are infected by COVID-19 asymptomatically, and in due course, they also develop the antibody to resist COVID-19. In this way, herd immunity can be achieved through community transmission.

Based on these symptoms and conditions, the risk condition of the patient can be deduced using our proposed approach. The risk or the infection status has been classified into three classes, which are Not Infected, Mildly Infected, and Severely Infected. The classes have been mapped to the words in the "RISK" column of the dataset as:
Not Infected = Low; Mildly Infected = Medium; Severely Infected = High

Case 1: Not Infected

Among all the symptoms that have been considered, certain symptoms occur quite often for any human being. For example, people with a common cold also experience a dry cough, but only these conditions are not enough to ascertain COVID-19 infection. Moreover, billions of people fall under the category of the current smoker, but without the presence of the other symptoms we can easily declare the person to be in the class of Not Infected.

Case 2: Mildly Infected

A person showcasing one or two symptoms simultaneously cannot be confirmed with COVID-19 but can progress to a critical stage if proper containment measures are not taken. Such a person falls under this class of mildly infected.

Case 3: Severely Infected

A person with more than 3–4 symptoms, each of which has crossed their normal limits, shows positive results for COVID-19 in the majority of cases. A person can be judged to be at high risk if they are suffering from high fever, high breathing rates, and are facing persistent chest pain.

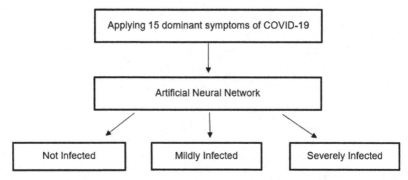

FIGURE 8.1   Block diagram of proposed COVID-19 prediction methodology.

The dataset is then passed to the ANN classifier. Due to the ability to reproduce and model nonlinear processes, ANN has been used keeping in mind the inconsistency of the data related to coronavirus. In section 8.5, the basic principles of ANN are described along with the proper justification of using this model in our proposed approach. Finally, ANN predicts the output among one of the three classes, that is, Not Infected, Mildly Infected, and Severely Infected (see Fig. 8.1).

8.5   Brief description of artificial neural networks

ANN, an integral part of machine learning (Biswas et al., 2021; Saha, Saha, Kundu, Chakraborty, & Banerjee, 2020; Banerjee et al., 2019; Pandey et al., 2019; Chattopadhyay et al., 2020), teaches the computer to think like a human. A computer can learn to execute classification jobs based on text, sound, or images. An ANN is trained by using a large dataset. Neural network architecture (Chakraborty & Banerjee, 2013; Chakraborty, Banerjee, & Chattopadhyay, 2017, 2019, 2020) contains many layers. ANN is an information processing system that is inspired by the way the nervous system processes information. It is formed of many highly interconnected processing elements (neurons) working together to solve a specific problem.

8.5.1   Principles of artificial neural network

To understand neural network, we need to discuss the following:

- Neurons—Biological neurons have inspired the development of the general model of ANN. Perceptron, which is a single layer of neural network, gives a single output.

In Fig. 8.2 for a single observation, $x_1, x_2, ...x_{(n)}$ describes independent variables to the network. Each of these inputs are multiplied by connection weight $\left(w_1, w_2, ...w_{(n)}\right)$. The strength of each node is depicted by the

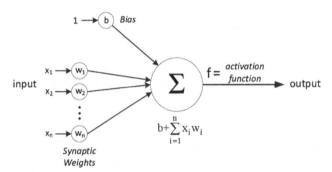

FIGURE 8.2 Perceptron.

respective weights. The activation function is shifted up or down through the bias value denoted by b.

These products are primarily added and fed to an activation function to produce a result in the form of output that is sent.

$$x_1.w_1 + x_2.w_2 + ...x_n.w_n = \sum x_i.w_i$$

The activation function is employed now, $\varphi(\sum x_i.w_i)$.

- Activation functions—An activation function is essential for the ANN. It helps to understand something very complicated. The main purpose is to process the input signals and convert them into the corresponding output signal. The next layer in the stack receives input from this output signal. The activation of the neuron depends upon the activation function. The decision is made by adding bias and calculating the weighted sum to it, which helps in the introduction of nonlinearity to the output signal coming out of a neuron. Without the activation function, the output signal would be a one-degree polynomial, that is, a linear function that is simple to solve but has the least power to solve complex problems.

Nonlinear functions have more than one degree and have a curvature.

"Universal function approximators," is the other name of neural network, which means it tends to learn and compute any function.

Different kinds of activation functions:

1. Binary step function or threshold activation function—It is a threshold-based activation. If the input value is above a certain threshold and sends the same signal to the next layer, then only the neuron is activated (see Fig. 8.3).

If $Y >$ threshold then the activation function A = "activated" otherwise not or another way, if $Y >$ threshold 0, then A = 1.

2. Logistic function or sigmoid activation function—Sigmoid curve is a mathematical function with a characteristic "S"-shaped curve in the range of 0 and 1. Therefore it is primarily used to prognosticate the probability of an output.

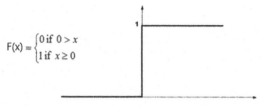

$$F(x) = \begin{cases} 0 \text{ if } 0 > x \\ 1 \text{ if } x \geq 0 \end{cases}$$

FIGURE 8.3   Binary step function.

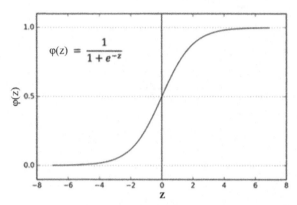

$$\varphi(z) = \frac{1}{1 + e^{-z}}$$

FIGURE 8.4   Sigmoid curve.

As we can see in Fig. 8.4, we can find the curve's slope at any two points.

We can infer from the above graph that:

- F(z) tends toward 1 as $z -> \infty$
- F(z) tends toward 0 as $z -> -\infty$
- F(z) is always bounded between 0 and 1

3. Hyperbolic tangent function (tan h)—It has similar characteristics to sigmoid, but it performs better. We can stack layers due to its nonlinear nature. The function ranges within $(-1, 1)$ (see Fig. 8.5).

This function maps the substantial negative inputs to negative outputs, and to near-zero outputs, only zero-valued inputs are mapped. Hence, during training, it is less likely to get stuck.

4. Rectified Linear Units (ReLu)—In conventional neural network and ANN, it is the most used activation function, ranging from zero to infinity.

If x is positive, then ReLu gives an output "x" and 0 otherwise. A combination of ReLu is also nonlinear as ReLu is nonlinear. Any function can be approximated with a combination of ReLu as it is a valid approximator (see Fig. 8.6).

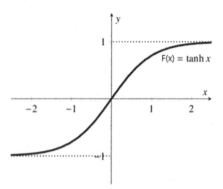

FIGURE 8.5 Hyperbolic tangent function.

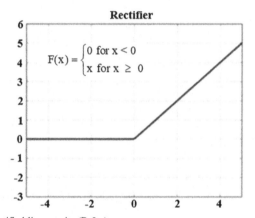

FIGURE 8.6 Rectified linear units (ReLu).

To the hidden layers of neural network, ReLu can only be applied. As the problem we are solving is a classification problem, we need to use a SoftMax function for the outer layer.

- How do neural networks work—The input values pass within weighted synapses to the output layer. All the input parameters are to be analyzed, an activation function is employed to it in the neuron, and then the result will be produced.

There is a way to improve the neural network's power and increase its accuracy by adding hidden layers that sit between output and input layers.

All the four variables are attached to neurons through a synapse; it is visible from Fig. 8.7. Still, all the synapses are not weighted; it is either a nonzero value or 0. The value 0 indicates that they will be rejected, and the nonzero value indicates importance. With different combinations of variables, many neurons do similar calculations. That is why neural network is so powerful.

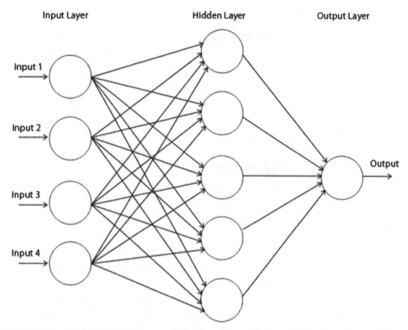

FIGURE 8.7 The hidden layer is shown within a neural network.

In a very flexible way, the neurons interact and work, allowing each to look for specific things. This approach allows them to do a comprehensive search for what it is trained for.

- How do neural networks learn—The cost function plays a vital role in training neural nets. For all the neural network layers, to adjust the weight and threshold of the next input, the cost function is analyzed. The cost function is one-half of the squared difference actual and output value. The lower the cost function, the more accurate is the prediction. Thus after each run, the error function keeps on getting lesser and lesser. As long as there is a mismatch between the actual value and predicted value, the weights are continuously adjusted during each run. After this adjustment, the neural network is run again, which forms a new cost function. This process is repeated until we make the cost function small and acceptable for the application-specific task.

Fig. 8.8 shows how back-propagation works. It is applied continuously through a network until the error value is kept at a minimum.

8.6 Parameter settings for the proposed ANN model

The major aspect of using the ANN is because they can learn by themselves and give the output that is not limited to the input data provided to the neural

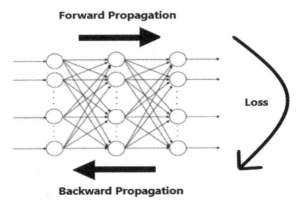

FIGURE 8.8   Back-propagation in action.

network. This network can learn from examples and apply them when a similar type of event arises. In the current situation of COVID-19 we need a model that can learn by itself.

For this study, we have split the dataset into 80% train sets and 20% test sets. The dataset contains numeric values for attributes like body temperature and breathing rate. Other than that, all other attributes have yes/no value. The age attribute has been categorized into three classes, which are mapped as $A = 0-20$ years, $B = 21-40$ years, and $C > 41$ years. The total number of inputs for each record is 15. As a result, the input layer of the network contains 15 nodes. This layer provides information from the outside world to the network. No computation is performed here. The output layer contains just a single node since we are predicting a single output among the three output classes. The number of nodes of the hidden layer is computed from the average of nodes of the input layer and output layer, which is 8. The hidden layers perform all the computations on the outside information brought by the input layer and transfer the result to the output layer.

An activation function is important for the ANN. It is responsible for making sense of something complicated. The main purpose is to process the input signal and convert them into the corresponding output signal. The purpose of the activation function is to introduce nonlinearity into the output of a neuron. The activation function calculates the weighted sum and decides whether a neuron should be activated or not and further adds bias with it. The purpose of the activation function is to introduce nonlinearity into the output of a neuron. We are using ReLu as the neural network's activation function. ReLu is six times improved over hyperbolic tangent function. It is applied to the hidden layers of the neural network. It gives an output "x" if x is positive and 0 otherwise. ReLu is a good approximator and any function can be approximated with a combination of ReLu. Since the dataset created for this study consists of a minimal record, the batch size that has been used for training the model has been kept as 10. As a result, for each iteration the algorithm takes 10 records to train the network and update the weights to minimize the error before proceeding with the next batch of 10 records. To improve the accuracy of the model the network has been trained for five iterations or epochs.

8.7 Experimental results and discussion

The dataset that has been created for this study contains 230 records, with 15 attributes that are described in section 8.3. Due to the unavailability of a proper dataset, authors are bound to employ the machine learning algorithm on a minimal number of data. The authors also believe that the proposed technique will perform better if provided with a real-time dataset. In this dataset (see Appendix, Table A1), the output column, that is, the RISK column, contains three different values, "Low," "Medium," and "High," which have been mapped to Not Infected, Mildly Infected, and Severely Infected, respectively.

Fig. 8.9 gives a clear visualization of the risk case distribution of COVID-19 based on our dataset. The authors have been very meticulous in preparing the dataset where the impact of various factors and conditions have been taken into consideration to assign a case as high risk. As a result of which a better accuracy has been achieved while predicting a patient to be at high risk. Keeping in mind the importance of a more accurate prediction of COVID-19 risk, ReLu has been used as an activation function that has been discussed in section 8.5. The ANN has been implemented by using python in Spyder platform.

From the confusion matrix in Fig. 8.10, we see that the number of correct predictions is quite high, which is depicted by the diagonal values in the matrix. The total number of correct predictions is 39 (diagonal values) and the total number of records tested is 46. Therefore the accuracy becomes $(39/46) \times 100 = 84.7$ (approximately).

In Table 8.1, the classifier on the test set has shown high precision for Not Infected class and moderate results for the Severely Infected class. This means that very few cases have been labeled wrong. Since all the predictions of Mildly Infected cases were labeled wrong, the precision obtained for that class is 0. However, with real-time data, the model is expected to function more efficiently and thus provide more accurate results. The classifier has shown perfect results for predicting the cases that belong to the Severely Infected class since it has recall = 1. This is again a very crucial aspect since the correct prediction of high risk gives our approach more reliability. The f1-score is simply the average of precision and recall. The final column "Support" gives the number of true values for each class in the test data.

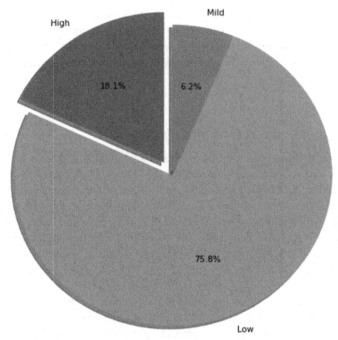

FIGURE 8.9   Distribution of risk cases of COVID-19.

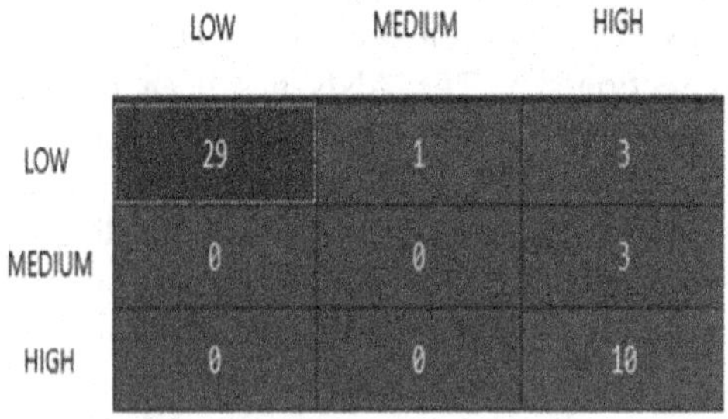

	LOW	MEDIUM	HIGH
LOW	29	1	3
MEDIUM	0	0	3
HIGH	0	0	10

FIGURE 8.10   Confusion matrix.

TABLE 8.1 Classification report.

Class	Precision	Recall	f1-score	Support
Not Infected	1.00	0.30	0.47	33
Mildly Infected	0.00	0.00	0.00	3
Severely Infected	0.56	1.00	0.71	10

8.8   Performance comparison between ANN and other classification algorithms

The authors in this chapter analyzed the COVID-19 dataset by predicting the risk factor using all the supervised models. Among all the other supervised models, ANN works best in predicting COVID-19 risk with maximum accuracy. To support our claim, we have performed a comparative study on the popular supervised learning models. Looking at the current pandemic scenario, it is visible how quickly the COVID-19 is spreading its roots and how its symptoms have evolved over the last few months since its outbreak. Therefore, the concrete data related to the symptoms will further increase the number of predictors for COVID-19 data analysis and predictions. Keeping all these in mind, we propose the usage of visual programming methodology using Orange (Maughan, 2019) to quickly analyze the data with a set of predictors and see their efficiency in predicting the disease.

Orange is an open source machine learning and data visualization toolkit. It not only facilitates easy analysis of data using machine learning models from researchers from any background but also reduces coding overheads. We have analyzed the performance comparison between different machine learning models using Orange. Figs. 8.11−8.13 depict the methodology implemented by Orange. Initially the COVID-19 dataset is parsed and based on the predictive score the features are ranked. This step ensures the selection of features that have the best correlation with the predictor. In general, the number of features gets reduced. Moreover, this "Rank" block helps us to identify the linearly varying redundant features and thus the overall computation time gets reduced (see Fig. 8.13). This phenomenon makes the methodology fast, scalable, and robust. Popular scoring methods (Orange Visual Programming, 2020) like Info Gain, FCBF, Gini, and ReliefF are employed for ranking purposes. After that the top features are selected and fed into the machine learning models for training and prediction. The results of the prediction are evaluated by the "Test and Score" module and also provides comparison metrics. These metrics give us a clear visualization of the

Evaluation Results

Model	AUC	CA	F1	Precision	Recall
kNN	0.980	0.977	0.976	0.976	0.977
Tree	0.968	0.968	0.967	0.967	0.968
Random Forest	0.987	0.949	0.946	0.944	0.949
Neural Network	0.991	0.949	0.945	0.944	0.949
Logistic Regression	0.977	0.945	0.935	0.936	0.945
AdaBoost	0.954	0.965	0.965	0.964	0.965

FIGURE 8.11   Evaluation results.

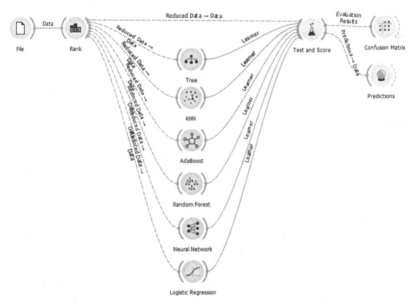

FIGURE 8.12 Schematics of visual programming.

	#	Info. gain	Gain ratio	Gini	ANOVA	χ^2	ReliefF	FCBF
Breathing Rate		0.671	0.340	0.265	NA	112.876	0.504	0.824
Chest Pain	2	0.569	0.678	0.231	NA	125.489	0.846	1.628
Tiredness	2	0.373	0.391	0.139	NA	64.484	0.617	0.620
Body Temp		0.356	0.178	0.136	NA	67.826	0.163	0.312
Hypertension	2	0.224	0.277	0.100	NA	58.102	0.565	0.330
Dry Cough	2	0.214	0.228	0.087	NA	43.378	0.612	0.284
Heart Problems	2	0.171	0.210	0.080	NA	44.297	0.177	0.232
Contact with a person with fever or cold in last few days	2	0.155	0.270	0.080	NA	50.895	0.158	0.246
Age	3	0.142	0.093	0.043	NA	18.752	0.528	0.127
Sore Throat	2	0.124	0.156	0.054	NA	32.874	0.242	0.160
Diabetes	2	0.115	0.179	0.058	NA	35.895	0.154	0.163
Nasal Congestion	2	0.083	0.101	0.036	NA	21.125	0.319	0.100
Diarrhea	2	0.082	0.102	0.033	NA	21.621	0.140	0.101
Runny Nose	2	0.069	0.117	0.033	NA	21.549	0.171	0.095
Current Smoker	2	0.066	0.081	0.030	NA	16.915	0.218	0.000

FIGURE 8.13 Rank of features.

performance of various supervised machine learning models viz. Tree, KNN, AdaBoost, Random Forest, Neural Network, and Logistic Regression. The evaluation results of the models are depicted in Fig. 8.10 and ANN outperforms all other models that are tested. We have done our prediction based on the parameters like Area under receiver operating characteristic (ROC), Classification Accuracy, F1 Score, Precision, and Recall. The confusion

(a) Tree

Actual \ Predicted	High	Low	Medium	Σ
High	126	4	0	130
Low	0	513	7	520
Medium	2	9	29	40
Σ	128	526	36	690

(b) KNN

Actual \ Predicted	High	Low	Medium	Σ
High	126	4	0	130
Low	0	518	2	520
Medium	6	4	30	40
Σ	132	526	32	690

(c) AdaBoost

Actual \ Predicted	High	Low	Medium	Σ
High	126	4	0	130
Low	0	518	2	520
Medium	6	4	30	40
Σ	132	526	32	690

(d) Random forest

Actual \ Predicted	High	Low	Medium	Σ
High	120	3	7	130
Low	0	517	3	520
Medium	16	6	18	40
Σ	136	526	28	690

(e) Neural network

Actual \ Predicted	High	Low	Medium	Σ
High	122	0	8	130
Low	0	517	3	520
Medium	20	4	16	40
Σ	142	521	27	690

(f) Logistic regression

Actual \ Predicted	High	Low	Medium	Σ
High	122	3	5	130
Low	0	520	0	520
Medium	22	8	10	40
Σ	144	531	15	690

FIGURE 8.14 Confusion matrix of respective supervised learning algorithms (A) Tree (B) KNN (C) AdaBoost (D) Random forest (E) Neural network (F) Logistic regression.

matrix for all the models has been summarized in Fig. 8.14, illustrating the superiority of ANN in predicting COVID-19 efficiently.

8.9 Conclusion

This chapter aims to develop a model that can predict whether a person is affected by COVID-19 or not using ANN. The symptoms or attributes and conditions have been selected for the dataset, keeping in mind the evolving nature of the virus and its uncertainty. The conditions considered as an attribute in the dataset are to give more accuracy in predicting and segregating the Not Infected class and the Mildly Infected class. The Not Infected class, for the time being, can be considered safe and cannot be said with certainty since persons without any signs and symptoms of COVID-19 are also getting affected. A person who has been predicted to be in the Mildly Infected class should undergo proper medication and containment measures to prevent this condition's elevation to a severe state. A person who is in the Severely Infected class should immediately undergo proper medication to overcome the disease. Authors have shown that ANN outperforms other supervised models for the purpose of predicting COVID-19. The accuracy obtained by implementing ANN by Python in the Spyder platform is 84.7% and in Orange platform is 99.1%. Due to the minimal size of the dataset and different platforms, the accuracy obtained is different in both cases. To reduce the margin of error concerning the platform predicting the result, the authors have considered 84.7% accuracy to predict COVID-19. Moreover, with each new entry of the data related to COVID-19 patients, the prediction accuracy improves. Such type of approach improves the results in better prediction and on-time treatment of the patient.

Appendix

TABLE A1 Dataset snapshot.

Age	Body temp	Dry cough	Tiredness	Chest pain	Nasal congestion	Runny nose	Sore throat	Diarrhea	Breathing	Hypertension	Cardiovascular Diseases	Diabetes	Current smoker	Contact with a person with fever	Risk
A	98.7	Yes	No	No	Yes	No	Yes	No	15	No	Yes	No	No	Yes	Low
A	100.2	No	No	No	No	No	No	No	16	No	No	No	No	No	Low
A	102.8	Yes	Yes	No	No	Yes	No	No	14	No	No	No	No	No	Low
A	99	Mo	No	No	No	No	No	No	18	No	No	No	No	No	Low
A	99.6	Yes	Yes	No	No	No	No	Yes	17	No	No	No	No	No	Low
A	101.1	No	No	No	No	No	No	No	16	No	No	No	No	No	Low
A	100.8	No	No	No	No	No	No	No	20	Yes	Yes	No	No	No	Low
A	99.2	Yes	No	No	Yes	No	No	No	25	No	No	No	No	No	Low
A	99.5	Yes	Yes	No	No	No	No	No	14	No	No	No	No	No	Low
A	98.4	No	No	No	No	No	Yes	Yes	16	No	No	No	No	No	Low
A	101.3	No	No	No	No	Yes	No	No	18	No	No	No	No	No	Low
A	99	Yes	Yes	No	No	No	No	No	17	No	No	No	No	No	Low
A	100.6	No	No	No	No	No	No	Yes	15	No	No	No	No	No	Low
A	98.1	Yes	No	No	No	No	No	No	13	No	No	No	No	No	Low
A	103	No	Yes	No	No	No	No	No	16	No	No	No	No	No	Low
A	99.7	Yes	No	No	No	No	No	No	18	No	No	No	No	No	Low
A	102.2	No	No	No	No	Yes	No	No	16	No	No	No	No	No	Low
A	98.9	Yes	No	No	No	No	No	No	24	No	No	No	No	No	Low
A	103.5	Yes	Yes	Yes	Yes	No	Yes	Yes	28	No	No	No	No	Yes	Medium

References

Banerjee, J. S., & Chakraborty, A. (2014). Modeling of Software Defined Radio Architecture and Cognitive Radio: The Next Generation Dynamic and Smart Spectrum Access Technology. In *Cognitive radio sensor networks: Applications, architectures, and challenges* (pp. 127–158). IGI Global.

Banerjee, J. S., & Chakraborty, A. (2015). Fundamentals of software defined radio and cooperative spectrum sensing: a step ahead of cognitive radio networks. *In Handbook of research on software-defined and cognitive radio technologies for dynamic spectrum management* (pp. 499–543). IGI Global.

Banerjee, J., Maiti, S., Chakraborty, S., Dutta, S., Chakraborty, A., & Banerjee, J.S. (2019). Impact of machine learning in various network security applications. In *2019 3rd International Conference on Computing Methodologies and Communication (ICCMC)*, (pp. 276–281). IEEE.

Banerjee, J. S., Chakraborty, A., & Chattopadhyay, A. (2017). *Fuzzy based relay selection for secondary transmission in cooperative cognitive radio networks. Advances in Optical Science and Engineering* (pp. 279–287). Singapore: Springer.

Banerjee, J. S., Chakraborty, A., & Chattopadhyay, A. (2018a). A novel best relay selection protocol for cooperative cognitive radio systems using fuzzy AHP. *Journal of Mechanics of Continua and Mathematical Sciences*, *13*(2), 72–87.

Banerjee, J. S., Chakraborty, A., & Chattopadhyay, A. (2018b). *Relay node selection using analytical hierarchy process (AHP) for secondary transmission in multi-user cooperative cognitive radio systems. Advances in Electronics, Communication and Computing* (pp. 745–754). Singapore: Springer.

Banerjee, J. S., Chakraborty, A., & Chattopadhyay, A. (2018c). Reliable best-relay selection for secondary transmission in co-operation based cognitive radio systems: A multi-criteria approach. *Journal of Mechanics of Continua and Mathematical Sciences*, *13*(2), 24–42.

Banerjee, J. S., Chakraborty, A., & Chattopadhyay, A. (2021). A decision model for selecting best reliable relay queue for cooperative relaying in cooperative cognitive radio networks: the extent analysis based fuzzy AHP solution. *Wireless Networks*, *27*(4), 2909–2930.

Banerjee, J. S., Chakraborty, A., & Karmakar, K. (2013). Architecture of cognitive radio networks. In *Cognitive radio technology applications for wireless and mobile ad hoc networks*, (pp. 125–152). IGI Global.

Banerjee, J. S., & Karmakar, K. (2012). A comparative study on cognitive radio implementation issues. *International Journal of Computer Applications*, *45*(15), 44–51.

Banerjee, J. S., Goswami, D., & Nandi, S. (2014). OPNET: a new paradigm for simulation of advanced communication systems. *In Proceedings of International Conference on Contemporary Challenges in Management, Technology & Social Sciences, SEMS* (pp. 319–328).

Baxt, W. G. (1991). Use of an artificial neural network for the diagnosis of myocardial infarction. *Annals of Internal Medicine*, *115*(11), 843–848.

Bhatnagar, V., Poonia, R. C., Nagar, P., Kumar, S., Singh, V., Raja, L., & Dass, P. (2020). Descriptive analysis of COVID-19 patients in the context of India. *Journal of Interdisciplinary Mathematics*, 1–16. Available from https://doi.org/10.1080/09720502.2020.1761635.

Biswas, S., Sharma, L. K., Ranjan, R., Saha, S., Chakraborty, A., & Banerjee, J. S. (2021). Smart farming & water saving based intelligent irrigation system implementation using the internet of things. *In Recent Trends in Computational Intelligence Enabled Research* (pp. 339–354). Academic Press, Elsevier.

Biswas, S., Sharma, L. K., Ranjan, R., & Banerjee, J. S. (2020). Go-COVID: An interactive crossplatform based dashboard for real-time tracking of COVID-19 using data analytics. *Journal of Mechanics of Continua and Mathematical Sciences, 15*(6), 1−15.

Chattopadhyay, J., Kundu, S., Chakraborty, A., Banerjee, J.S. (2020). Facial expression recognition for human computer interaction. In *International Conference on Computational Vision and Bio Inspired Computing*, (pp. 1181−1192). Springer, Cham.

Chakraborty, A., & Banerjee, J. S. (2013). An advance Q learning (AQL) approach for path planning and obstacle avoidance of a mobile robot. *International Journal of Intelligent Mechatronics and Robotics (IJIMR), 3*(1), 53−73.

Chakraborty, A., Banerjee, J. S., & Chattopadhyay, A. (2017). Non-uniform quantized data fusion rule alleviating control channel overhead for cooperative spectrum sensing in cognitive radio networks. *In 2017 IEEE 7th International Advance Computing Conference (IACC)* (pp. 210−215). IEEE.

Chakraborty, A., Banerjee, J. S., & Chattopadhyay, A. (2019). Non-uniform quantized data fusion rule for data rate saving and reducing control channel overhead for cooperative spectrum sensing in cognitive radio networks. *Wireless Personal Communications, 104*(2), 837−851.

Chakraborty, A., Banerjee, J. S., & Chattopadhyay, A. (2020). Malicious node restricted quantized data fusion scheme for trustworthy spectrum sensing in cognitive radio networks. *J. Mech. Contin. Math. Sci, 15*(1), 39−56.

Chen, N., Zhou, M., Dong, X., Qu, J., Gong, F., Han, Y., & Yu, T. (2020). Epidemiological and clinical characteristics of 99 cases of 2019 novel coronavirus pneumonia in Wuhan, China: A descriptive study. *The Lancet, 395*(10223), 507−513.

Das, D., Pandey, I., Chakraborty, A., & Banerjee, J. S. (2017). Analysis of implementation factors of 3D printer: the key enabling technology for making prototypes of the engineering design and manufacturing. *International Journal of Computer Applications, 975*, 8887.

Das, D., Pandey, I., & Banerjee, J. S. (2016). An in-depth study of implementation issues of 3D printer. In *Proceedings of MICRO 2016 Conference on Microelectronics, Circuits and Systems* (pp. 45-49).

Fang, L., Karakiulakis, G., & Roth, M. (2020). Are patients with hypertension and diabetes mellitus at increased risk for COVID-19 infection? *The Lancet. Respiratory Medicine.*

Guan, W. J., Ni, Z. Y., Hu, Y., Liang, W. H., Ou, C. Q., He, J. X., & Du, B. (2020). Clinical characteristics of coronavirus disease 2019 in China. *New England Journal of Medicine, 382* (18), 1708−1720.

Guhathakurata, S., Kundu, S., Chakraborty, A., & Banerjee, J. S. (2020). A novel approach to predict COVID-19 using support vector machine. In *Data Science for COVID-19*, (pp. 351−364). Elsevier.

Guhathakurata, S., Saha, S., Kundu, S., Chakraborty, A., & Banerjee, J. S. (2020a). *South Asian countries are less fatal concerning COVID-19: A hybrid approach using machine learning and M-AHP. Computational Intelligence Techniques for Combating COVID-19.* Springer (press).

Guhathakurata, S., Saha, S., Kundu, S., Chakraborty, A., & Banerjee, J. S. (2020b). South Asian countries are less fatal concerning COVID-19: A fact-finding procedure integrating machine learning & Multiple Criteria Decision Making (MCDM) technique. *Journal of the Institution of Engineers (India): Series B*, Springer (Communicated).

Guresen, E., Kayakutlu, G., & Daim, T. U. (2011). Using artificial neural network models in stock market index prediction. *Expert Systems with Applications, 38*(8), 10389−10397.

Hill, T., Marquez, L., O'Connor, M., & Remus, W. (1994). Artificial neural network models for forecasting and decision making. *International Journal of Forecasting, 10*(1), 5−15.

Hsu, K. L., Gupta, H. V., & Sorooshian, S. (1995). Artificial neural network modeling of the rainfall-runoff process. *Water Resources Research, 31*(10), 2517−2530.

Huang, C., Wang, Y., Li, X., Ren, L., Zhao, J., Hu, Y., & Cheng, Z. (2020). Clinical features of patients infected with 2019 novel coronavirus in Wuhan, China. *The Lancet, 395*(10223), 497−506.

Huang, S., Shen, Q., & Duong, T. Q. (2010). Artificial neural network prediction of ischemic tissue fate in acute stroke imaging. *Journal of Cerebral Blood Flow & Metabolism, 30*(9), 1661−1670.

Jiang, J., Trundle, P., & Ren, J. (2010). Medical image analysis with artificial neural networks. *Computerized Medical Imaging and Graphics, 34*(8), 617−631.

Kumar, M., Raghuwanshi, N. S., Singh, R., Wallender, W. W., & Pruitt, W. O. (2002). Estimating evapotranspiration using artificial neural network. *Journal of Irrigation and Drainage Engineering, 128*(4), 224−233.

Kumari, R., Kumar, S., Poonia, R. C., Singh, V., Raja, L., Bhatnagar, V., & Agarwal, P. (2020). Analysis and predictions of spread, recovery, and dDeath caused by COVID-19 in India. *Big Data Mining and Analytics, IEEE*. Available from https://doi.org/10.26599/BDMA.2020.9020013.

Lee, K. Y., Cha, Y. T., & Park, J. H. (1992). Short-term load forecasting using an artificial neural network. *IEEE Transactions on Power Systems, 7*(1), 124−132.

Maughan, J. Machine Learning with Orange, (2019). https://medium.com/@jackmaughan_50251/machine-learning-with-orange-8bc1a541a1d7/ (accessed August 10, 2020).

Novel, C.P.E.R.E. (2020). The epidemiological characteristics of an outbreak of 2019 novel coronavirus diseases (COVID-19) in China. *Zhonghua liu xing bing xue za zhi = Zhonghua liuxingbingxue zazhi, 41*(2), 145.

Orange Visual Programming (2020). https://orange-visual-programming.readthedocs.io/widgets/data/rank.html/ (accessed August 10, 2020).

Pandey, I., Dutta, H.S., & Banerjee, J.S. (2019). WBAN: A smart approach to next generation e-healthcare system. In *2019 3rd International Conference on Computing Methodologies and Communication (ICCMC)*, (pp. 344−349). IEEE.

Paul, S., Chakraborty, A., & Banerjee, J.S. (2017). A fuzzy AHP-based relay node selection protocol for wireless body area networks (WBAN). In *2017 4th International Conference on Opto-Electronics and Applied Optics (Optronix)*, (pp. 1−6). IEEE.

Paul, S., Chakraborty, A., & Banerjee, J. S. (2019). *The extent analysis based fuzzy AHP approach for relay selection in WBAN. Cognitive Informatics and Soft Computing* (pp. 331−341). Singapore: Springer.

Rothan, H. A., & Byrareddy, S. N. (2020). The epidemiology and pathogenesis of coronavirus disease (COVID-19) outbreak. *Journal of Autoimmunity, 102433.*

Roy, R., Dutta, S., Biswas, S., & Banerjee, J.S., *Android things: A comprehensive solution from things to smart display and speaker.* In *Proceedings of International Conference on IoT Inclusive Life (ICIIL 2019), NITTTRChandigarh, India* Singapore (pp. 339−352) Springer.

Saha, O., Chakraborty, A., & Banerjee, J.S. (2017). A decision framework of IT-based stream selection using analytical hierarchy process (AHP) for admission in technical institutions. In *2017 4th International Conference on Opto-electronics and Applied Optics (Optronix)*, (pp. 1−6). IEEE.

Saha, O., Chakraborty, A., & Banerjee, J. S. (2019). *A fuzzy AHP approach to IT-based stream selection for admission in technical institutions in India. Emerging Technologies in Data Mining and Information Security* (pp. 847−858). Singapore: Springer.

Saha, P., Saha, S., Kundu, S., Chakraborty, A., & Banerjee, J. S. (2020). *Application of machine learning in app-based cab booking system: A survey on Indian scenario. Applications of Artificial Intelligence in Engineering.* Springer (Press).

Singh, V., Poonia, R. C., Kumar, S., Dass, P., Agarwal, P., Bhatnagar, V., & Raja, L. (2020). Prediction of COVID-19 coronavirus pandemic based on time series data using support vector machine. *Journal of Discrete Mathematical Sciences & Cryptography*. Available from https://doi.org/10.1080/09720529.2020.1784525.

Sinha, S. K., & Wang, M. C. (2008). Artificial neural network prediction models for soil compaction and permeability. *Geotechnical and Geological Engineering, 26*(1), 47−64.

U.S. Centers for Disease Control and Prevention (CDC), (2020). How COVID-19 spreads. Retrieved August 6, 2020.

United States Centers for Disease Control and Prevention (CDC), (2020a). Interim Clinical Guidance for Management of Patients with Confirmed Coronavirus Disease (COVID-19). August 4, 2020.

United States Centers for Disease Control and Prevention (CDC), (2020b). Symptoms of Novel Coronavirus (2019-nCoV). Retrieved February 10, 2020.

Velavan, T. P., & Meyer, C. G. (2020). The COVID-19 epidemic. *Tropical Medicine & International Health, 25*(3), 278−280. Available from https://doi.org/10.1111/tmi.13383. PMC7169770, PMID 32052514.

Wang, W., Tang, J., & Wei, F. (2020). Updated understanding of the outbreak of 2019 novel coronavirus (2019-nCoV) in Wuhan, China. *Journal of Medical Virology, 92*(4), 441−447.

Webmd, (2020). https://www.webmd.com/lung/covid-19-symptoms#1 (Retrieved on: September 23, 2020).

WHO − China Joint Mission, (2020). Report of the WHO − China Joint Mission on Coronavirus Disease 2019 (COVID-19). https://www.who.int/docs/default-source/coronaviruse/who-china-joint-mission-on-covid-19-finalreport.pdf (accessed March 1, 2020).

Wikipedia, (2020). https://en.wikipedia.org/wiki/COVID-19_pandemic (Retrieved on: September 12, 2020).

World Health Organization, (2020a). Coronavirus disease 2019 (COVID-19) Situation Report—47. https://www.who.int/docs/default-source/coronaviruse/situation-reports/20200307-sitrep-47-covid19.pdf?sfvrsn = 27c364a4_2 (accessed March 7, 2020).

World Health Organization, (2020b). Coronavirus disease 2019 (COVID-19) Situation Report—47. https://www.who.int/docs/default-source/coronaviruse/situation-reports/20200307-sitrep-47-covid19.pdf?sfvrsn = 27c364a4_2 (accessed August 6, 2020).

Zheng, Y. Y., Ma, Y. T., Zhang, J. Y., & Xie, X. (2020). COVID-19 and the cardiovascular system. *Nature Reviews Cardiology, 17*(5), 259−260.

Zu, Z. Y., Jiang, M. D., Xu, P. P., Chen, W., Ni, Q. Q., Lu, G. M., & Zhang, L. J. (2020). Coronavirus disease 2019 (COVID-19): A perspective from China. *Radiology*, 200490.

Chapter 9

Rapid medical guideline systems for COVID-19 using database-centric modeling and validation of cyber-physical systems

Mani Padmanabhan

Faculty of Computer Applications, SSL, Vellore Institute of Technology, Vellore, India

9.1 Introduction

The massive damage on humanity has recorded in the history of a global pandemic; Spanish flu Ebola, Asian flu, severe acute respiratory syndrome (SARS) are different pandemics in the past decade, but COVID-19 has rolled out the entire world as of August, 2020 with 20,730,456 people infected, and 75,1154 losing their lives. This number creates mental health problems for normal human beings with stress, worry, lack of sleep, and emotional issues. As of August, 2020, there has been no proper vaccination found for this virus. More investigation are seriously working to find the vaccination but virus spread as fast among people currently the challenges amplified in rural areas. Due to the unavailability of COVID-19 medical experts, a huge spread in rural places (Hadi, Kadhom, Hairunisa, Yousif, & Mohammed, 2020). To overcome this problem, database-centric medical guideline systems are mandatory to avoid mental stress in an average person.

A database-centric approach in the medical guideline system was already implemented in 2008 (Islam, Hasan, Wang, Germack, & Noor-E-Alam, 2018). The analysis of European research projects in the current year shows most of the artificial intelligence-based medical guidelines systems are involved in the medical sector. Rapid medical guidelines systems based on cyber-physical systems with proper modeling, simulation, and validation are necessary during the COVID-19 pandemic.

Cyber-Physical Systems. DOI: https://doi.org/10.1016/B978-0-12-824557-6.00012-1

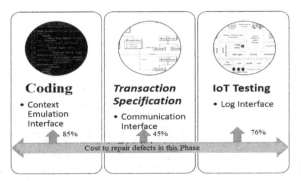

FIGURE 9.1 Cost expenditure percentage during re-engineering.

Rapid medical guidelines system based on cyber-physical systems can be more eye-catching among software developers (Guo, Fu, Zhang, Ren, & Sha, 2019). The cyber-physical courses, such as the human − machine interaction process during a global pandemic, are challenging to identify the preconditions during the development (Fox, Johns, & Rahmanzadeh, 1998).

Re-engineering is essential for the rapid medical guidelines system. Validation has assured the quality of products during the fast medical guidelines system (Guo, Fu, Zhang, Ren, & Sha, 2020; Shah, D. et al., 2019). The modeling and validation techniques are controlled by a simulation invocation relationship. Fig. 9.1 describes the percentage of cost spent on the dedication of defects in the different phases of rapid medical system development. The proposed simulation invocation algorithm identified the simulations to be affected during the human − machine interaction process. A medical professional-based methodology aims to improve the effectiveness of the product and reduce the development cost. To solve this problem, a stack allocation-based framework was proposed that utilizes two sources of information, which are code coverage and test execution time.

This rapid medical system helps rural medical officers and normal human beings to avoid unnecessary mental stress. The following sections describe the global pandemic COVID-19, framework for database-centric cyber-physical design, modeling, and validation techniques for rapid medical systems in detail.

9.2 Global pandemic of COVID-19

In late 2019 a novel type of coronavirus was identified as 2019-NCOV. COVID-19 arised in Wuhan, China. During March, 2020, the COVID-19 was declared as a global pandemic. In the past two decades, after SARS, yellow fever, cholera, and Ebola virus disease, respiratory coronavirus was detected in humans called COVID-19. It belongs to the family of coronaviruses, roniviridae, arteriviridae. The family is divided into alpha, beta, and gamma (Hadi et al., 2020). The pandemic in December 2019 was introduced with new cases in China. According to the laboratory investigation and earnest of the report finally recognized the virus were classified as COVID-19. On February 11, 2020, the World Health Organization stated the most famous name of the viruses as COVID-19.

The origin of COVID-19 in Wuhan, few adults went to local hospitals with severe respiratory symptoms for an unknown reason in November 2019. After SARS diseases spreading the control system and patient respiratory samples were sent to the references lab to discover the cases. The virus was recognized about 78% of similarity to SARS samples that were taken from the connectivity to human-related to the seafood markets were also positive, indicating that this virus was generated from seafood. The population of Wuhan around 11 million they were restricted with the public transaction with even though on January 2020 other cases appeared in Japan, South Korea, and Thailand on February 26, 2020, the confirmed cases according to the World Health Organization coronavirus database 56,650. In the first week of August 2020, around confirmed death is 731,641 were 19,905,163 cases worldwide Fig. 9.2.

The COVID-19 is confirmed in the 215 global countries as per the source from World Health Organization on August 15, 2020. COVID-19 pandemic is having a disproportionate impact on their livelihoods around all the sectors (Guo et al., 2016). Tourism revenue loss between January and May is "more than three times the loss during the Global Financial Crisis of 2009" reported by the World Tourism Organization. Millions of smallholder family farmers in Asia-Pacific produce a majority of the world's food. In Asia-Pacific, smallholder farmers own and operate the majority of farmland. The production of 75% has sold on to markets. The lockdown during this global pandemic restrictions in markets to curb the spread of COVID-19 the smallholder farmers have shattered the revenue.

COVID-19 pandemic is having a disproportionate impact on their livelihoods around all the sectors (Guo et al., 2016). Tourism revenue loss

Distribution of cases

India: 11.83 % (2,527,308 cases)
Russia: 4.27 % (912,823 cases)
South Africa: 2.71 % (579,140 cases)
Peru: 2.42 % (516,296 cases)
Brazil: 15.35 % (3,278,895 cases)
Mexico: 2.39 % (511,369 cases)
Colombia: 2.08 % (445,111 cases)
Chile: 1.79 % (382,111 cases)
Spain: 1.68 % (358,843 cases)
Iran: 1.59 % (338,825 cases)
United Kingdom: 1.48 % (316,367 cases)
United States: 25.64 % (5,476,266 cases)

FIGURE 9.2   Globally first 100 days confirmed cases of COVID-19 reported to WHO.

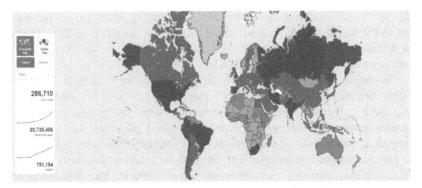

FIGURE 9.3 Global pandemic of COVID-19. *World Health Organization as on August 15, 2020.*

between January and May is "more than three times the loss during the Global Financial Crisis of 2009," reported by the World Tourism Organization. Millions of smallholder family farmers in Asia-Pacific produce a majority of the world's food. In Asia-Pacific, smallholder farmers own and operate the majority of farmland. The production of 75% has sold on to markets. The lockdown during this global pandemic restrictions in markets to curb the spread of COVID-19 the smallholder farmers have shattered the revenue (Fig. 9.3).

The survey had 1556 respondents working professionals in the 22–45 age bracket, from five major world cities as layoffs and pay cuts in the wake of the pandemic have impacted people's cash flows. The production company's history had never shown zero production and zero sales, but due to the global pandemic of COVID-19, it was an unprecedented quarter result in a lockdown. To avoid the economic slowdown, rapid medical guideline systems for COVID-19 using database-centric cyber-physical systems are necessary for hospitals. The rapid system needs to be developed with less time and error-free during the COVID-19 pandemic. Effective modeling and validation will produce the errorless system in the medical sector for human interaction.

9.3 Database-centric cyber-physical systems for COVID-19

9.3.1 Cyber-physical systems

The availability of health facilities for humans is the fundamental human right in the world (Kim et al., 2010). As reported in the last section, the global pandemic of COVID-19 has challenged the health facilities such as medical specialists, healthcare components, and druggists to provide health services to humans. The huge demand for professionals in the medical field. Over the past several years, many healthcare applications have been developed to enhance the healthcare industry (McKinley et al., 2011). Recent improvements in information technology have revolutionized electronic healthcare research and industry (Rahmaniheris, Wu, Sha, & Berlin, 2016). Database-centric and artificial intelligence are currently two of the most important and trending pieces for innovation and predictive analytics in healthcare, leading the digital healthcare transformation (Shah, AD et al., 2019). The benefits of using this new data platform for community and population health include better health-care outcomes, improvement of clinical operations, reducing costs of care, and generation of accurate medical information.

The benefits of using this database-centric cyber-physical system for public community provides the better healthcare outcomes during the global pandemic, improvement of clinical operations, reducing costs of care, and generation of accurate medical information are the huge demand in rural places of the world. "Every village in the country would be connected with the optical network within 1000 days or less than three years," Prime Minister Narendra Modi said at the ramparts of Red Fort on India's 74th Independence Day. The demand in the rural places for COVID-19 medical professionals is reduced if database-centric cyber-physical systems are implemented in the rural places worldwide.

9.3.2   Flow of rapid database-centric cyber-physical system

Different wearable sensors are connected between database-centric cyber-physical systems and target humans (Xiao, Wang, lin, & Yu, 2008). Database from humans through an extensive-scale network worldwide provides an efficient and effective way of data collection, processing with medical professionals, and generating the solution during the global pandemic of COVID-19. The COVID-19 healthcare system needs the cyber-physical system-based decision-marking applications, clinical algorithms, databased centric of patient's information, and digital pharmaceuticals applications. Data-centric standardization and communication protocols can enable the medical system to deliver efficient healthcare services. Several stakeholders, such as patients, hospitals, and pharmacies, have accessed and securely maintained health records with less energy. Fig. 9.4 shows the flow of rapid database-centric cyber-physical systems for COVID-19. The rapid medical system for COVID-19 is inbuilt in professionals for monitoring the applications and patient information. The server connects the database-centric with smart contracts. The storage collects the data from sensing devices from the healthcare system—the sample interaction between the proposed devices, as described in the following section.

These incidents of system failures in database-centric cyber-physical systems indicate an inarguable fact that unspecified assumptions are dangerous and can lead to catastrophes. However, system developers continuously make assumptions about the interpretation of requirements, design decisions,

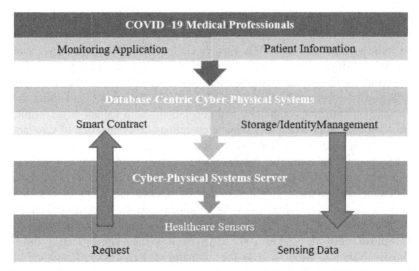

FIGURE 9.4 The flow of rapid database-centric cyber-physical systems for COVID-19.

the operational environment, characteristics of input data, and other factors during system development and deployment phases (Andrade and Machado, 2013). But these assumptions are seldom documented and less frequently validated by the domain experts who know to verify their appropriateness. Once the operating environment of cyber-physical systems violates the unspecified beliefs, failures may occur, and catastrophic accidents may happen, resulting in loss of revenue in the tune of hundreds of millions of dollars and loss of lives. A cyber-physical system for COVID-19, its execution behaviors are often impacted by its operating environment. However, the assumptions about a cyber-physical system's expected environment are often informally documented, or even left unspecified in the system design. Unfortunately, such anonymous system design environment assumptions made in safety-critical cyber-physical systems, such as medical cyber-physical systems performance as the medical professional in the world's rural places during the massive demand in the global pandemic.

9.4 Modeling and validation of rapid medical guideline systems

Software modeling and validation is significant for all the software development processes (Padmanabhan and Prasanna, 2017b). Rapid medical guideline systems are essential during COVID-19 for perfect testing because failures of software can cause serious damage. An unified modeling language

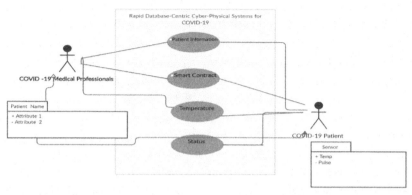

FIGURE 9.5 Modeling and validation of rapid database-centric cyber-physical systems.

sequence diagram has been recycled. Fig. 9.5 shows the process diagram with the binary value for the interaction between the human and cyber-physical systems.

The approach determines the goal of the patient interaction with medical professionals to find more important and effective potential data to find faults with less execution time with more code coverage. To support this goal, the proposed algorithm should find a stack simulation with a higher gain value and visit the stack at least once during the allocation (Padmanabhan and Prasanna, 2016; Padmanabhan and Prasanna, 2017a; Padmanabhan, 2018; Padmanabhan, 2019; Padmanabhan, 2020; Wang et al., 2018).

The Jaccard similarity methodology are well suitable for finding textually similar documents (Bingquan et al., 2018). To find the duplicate request in the stack flows, the minhash value to set S and T is |S U T|/|S ∩ T |, that is the ratio of the size of S and T's intersection the size of their union. The Jaccard similarity of S and T is to be represented by sim(S, T) (Wang et al., 2018). Textual similarity has importance during the huge demand worldwide for medical professionals. The Jaccard similarity addresses well about finding textually similar documents in an extensive system interaction in web or social media (Bhatnagar et al., 2020). The textual similarity has importance during the validation. This proposed technique reduced the cost and time during the chatbot testing.

Table 9.1 shows the list of sample interaction in the process rapid database-centric cyber-physical systems for COVID-19. The condition checking if the data are unavailable then produces the fallback message to the user. During the human discussion, the duplicate dialog flows have been counted to reduce the time during medical professional's interaction.

The proposed test path similarity identification methodology based on the minhashing matrix representation of collaborative modeling has yielded significant results in the rapid medical systems for COVID-19 interaction.

TABLE 9.1 List of sample requests to rapid medical systems for COVID-19.

Request ID	Sample source (sets S and T)	Jaccard similarity ID [sim(S, T)]	minhash value | S U T|/|S ∩ T|
R1	I am fine	J1	1
R2	I am well or not?	J1	0
R3	When will you discharge me?	J3	1
R4	I am in critical?	J1	0
R5	Tell my body temperature?	J4	0
R6	My last COVID-19 results shows positive or negative?	J5	1

9.5 Conclusion

The proposed rapid medical systems for COVID-19 allow us to address the health challenges and better understand patients' health during the global demand in medical professionals. It may not always be feasible for domain experts to identify the most safety-critical ones at different system development phases. Validation is mandatory in all the subdivisions of the system. Therefore, the proposed methodology uses the similarity index and database-centric cyber-physical technology. Software development steps with the validation technique of the cyber-physical system for COVID-19 address 99.5% issues in the world's rural places. In particular, the system development process with validation and verification in the design level and the development of rapid medical guideline systems for COVID-19 has a live connection with database-centric modeling and COVID-19 medical professionals. Thus the medical professional's study-based design of the medical guidelines system for COVID-19 has reduced development time, cost of deployment, testing, and maintenance based on the sample verification flow. More future work needs to be done on the platform to continue improving all the benefits for the entire health organization. Tools for performing the knowledge discovery process will be added to the rapid database-centric cyber-physical systems.

References

Andrade, W. L., & Machado, P. D. L. (2013). Generating test cases for real-time systems based on symbolic models. *IEEE Transactions on Software Engineering*, *39*(9), 1216–1229. Available from https://doi.org/10.1109/TSE.2013.13.

Bhatnagar, V., Poonia, R. C., Nagar, P., Kumar, S., Singh, V., Raja, L., & Dass, P. (2020). Descriptive analysis of COVID-19 patients in the context of India. *Journal of Interdisciplinary Mathematics*, 1–16.

Bingquan, L., et al. (2018). Content-oriented user modeling for personalized response ranking in chatbots. *IEEE/ACM Transactions on Audio, Speech, and Language Processing*, 26(1)).

Fox, J., Johns, N., & Rahmanzadeh, A. (1998). Disseminating medical knowledge: The PROforma approach. *Artificial Intelligence in Medicine*, 14(1–2), 157–182. Available from https://doi.org/10.1016/S0933-3657(98)00021-9.

Guo C., S. Ren, Y. Jiang, P.-L. Wu, L. Sha, & R.B. Berlin, Transforming medical best practice guidelines to executable and verifiable statechart models, in 2016 ACM/IEEE 7th International Conference on Cyber-physical Systems (ICCPS), Vienna, Austria, April 2016, pp. 1–10, Available from https://doi.org/10.1109/ICCPS.2016.7479121.

Guo, C., Fu, Z., Zhang, Z., Ren, S., & Sha, L. (2020). A framework for supporting the development of verifiably safe medical best practice guideline systems. *Journal of Systems Architecture*, 104, 101693. Available from https://doi.org/10.1016/j.sysarc.2019.101693, Mar.

Guo, C., Fu, Z., Zhang, Z., Ren, S., & Sha, L. (2019). Design verifiably correct model patterns to facilitate modeling medical best practice guidelines with statecharts. *IEEE Internet of Things Journal*, 6(4), 6276–6284. Available from https://doi.org/10.1109/JIOT.2018.2879475.

Hadi, A., Kadhom, M., Hairunisa, N., Yousif, E., & Mohammed, S. (2020). A review on COVID-19: Origin, spread, symptoms, treatment, and prevention. *Biointerface Research in Applied Chemistry*. Available from https://doi.org/10.33263/BRIAC106.72347242.

Islam, M., Hasan, M., Wang, X., Germack, H., & Noor-E-Alam, M. (2018). A systematic review on healthcare analytics: Application and theoretical perspective of data mining. *Healthcare*, 6(2), 54. Available from https://doi.org/10.3390/healthcare6020054.

Kim C., M. Sun, S. Mohan, H. Yun, L. Sha, & T.F. Abdelzaher, A framework for the safe interoperability of medical devices in the presence of network failures, in Proceedings of the 1st ACM/IEEE International Conference on Cyber-Physical Systems—ICCPS '10, Stockholm, Sweden, 2010, p. 149, Available from https://doi.org/10.1145/1795194.1795215.

McKinley, B. A., et al. (2011). Computer protocol facilitates evidence-based care of sepsis in the surgical intensive care unit. *The Journal of Trauma: Injury, Infection, and Critical Care*, 70(5), 1153–1167. Available from https://doi.org/10.1097/TA.0b013e31821598e9.

Padmanabhan, M., & Prasanna, M. (2017a). Test case generation for embedded system software using UML interaction diagram. *Journal of Engineering Science and Technology*, 12(4), 860–874.

Padmanabhan, M., & Prasanna, M. (2017b). Validation of automated test cases with specification path. *Journal of Statistics and Management Systems, Issue on Machine Learning and Software Systems*, 20(4), 535–542.

Padmanabhan M., 2018 A study on transaction specification based software testing for internet of things, in IEEE International Conference on Current Trends towards Converging Technologies (ICCTCT 2018), March1–3, 2018, Coimbatore, India.

Padmanabhan, M. (2019). Sustainable test path generation for chatbots using customized response. *International Journal of Engineering and Advanced Technology*, 8(6), 149–155. Available from https://doi.org/10.35940/ijeat.D6515.088619.

Padmanabhan, M. (2020). Test path identification for virtual assistants based on a chatbot flow specifications. In K. N. Das, J. C. Bansal, K. Deep, A. K. Nagar, P. Pathipooranam, & R. C. Naidu (Eds.), *Soft Computing for Problem Solving* (Vol. 1057, pp. 913–925). Singapore: Springer Singapore.

Padmanabhan M., Prasanna M., Automatic test case generation for programmable logic controller using function block diagram. IEEE International Conference on Information Communication and Embedded System (ICICES 2016), Chennai, India, ISBN: 978−1−5090−2552-7, 2016.

Rahmaniheris M., P. Wu, L. Sha, and R.R. Berlin, An Organ-centric best practice assist system for acute care, in 2016 IEEE 29th International Symposium on Computer-based Medical Systems (CBMS), Belfast and Dublin, Ireland, June 2016, pp. 100−105, doi: 10.1109/CBMS.2016.12.

Shah, A. D., et al. (2019). Recording problems and diagnoses in clinical care: Developing guidance for healthcare professionals and system designers. *BMJ Health & Care Informatics, 26* (1), e100106. Available from https://doi.org/10.1136/bmjhci-2019-100106.

Shah, D., et al. (2019). Recording problems and diagnoses in clinical care: Developing guidance for healthcare professionals and system designers. *BMJ Health & Care Informatics, 26*(1), e100106. Available from https://doi.org/10.1136/bmjhci-2019-100106.

Wang, H., et al. (2018). Social Media-based conversational agents for health management and interventions. *Computers, 51*, 26−33, August.

Xiao C., W. Wang, X. lin, and J.X. Yu, Efficient similarity joins for near duplicate detection, Proc. WWW Conference (2008), pp.131−140.

Chapter 10

Machine learning and security in Cyber Physical Systems

Neha V. Sharma[1], Narendra Singh Yadav[1] and Saurabh Sharma[2]
[1]Manipal University Jaipur, Jaipur, India, [2]Amity School of Hospitality, Amity University Jaipur, Jaipur, India

10.1 Introduction

Cyber Physical System (CPS) includes the coordination of the cyber world what is more, the physical world. Organized CPSs, for example, advanced mechanics systems, auto-pilot avionics, medical monitoring, process control systems, smart grids, Wireless Sensor Organizations (WSNs), and so on are exposed to serious examination and advancement. Endeavors are in progress to fabricate different specific reason CPSs, without any focused exertion to enhance hypothetical standards needed in their investigation and plan. In the previous years, many research scholars started to strengthen their research endeavors to show and examine unique reason to work on organized CPSs. At the point when CPSs advance on numerous time scales, synchronizing such frameworks is a significant issue. The writing on dynamical frameworks shows that the displaying of simultaneous frameworks got restricted intrigue. Machine learning converts the information into choices and activities quicker and absolutely. Machine learning strategy utilizes the information for expressive reason (to dissect what occurred), indicative reason (to consider for what reason did it occur), forecast (portraying about what will occur in future), and furthermore for prescriptive reason (which incorporates choice help and choice computerization). Machine learning helps to make expectations (which can change) when it is presented to new information. This information can be used to make a strong framework with the help of machine learning. Along these lines, we can believe that this enormous information and machine learning compliment one another. With the creation of CPS technology, the pace of development of information being created is exponential. This information must be utilized in a proficient way. There are two fundamental sorts of machine learning: supervised and unsupervised learning. Supervised learning depends on valuable data in named

Cyber-Physical Systems. DOI: https://doi.org/10.1016/B978-0-12-824557-6.00015-7
171

information. Characterization is the most well-known assignment in regulated learning; be that as it may, naming information physically is costly and tedious. Thus, the absence of adequate marked information shapes the fundamental bottleneck to administered learning. Unsupervised learning removes important element data from unlabeled information, making it a lot simpler to get preparing information. The discovery execution of unsupervised learning strategies is generally subpar compared to those of supervised learning techniques. Two more techniques belonging to machine learning are: semisupervised learning and reinforcement learning. Semisupervised learning is utilized when a mix of named and unlabeled information is available. It uses the benefit of supervised and unsupervised techniques. Reinforced learning is a natural-driven methodology that primarily uses trial and error strategy for learning. Dynamic learning is a subclass of this sort where client is a dynamic member in the learning process. The principle objective is to improve the model quality by procuring information from clients (Abu-Nimeh et al., 2007).

As seen in Fig. 10.1 machine learning strategies have been utilized in numerous regions of science in view of unique qualities like versatility and adaptability. A portion of the use of Artificial Intelligence incorporate climate forecasting, image processing (like face acknowledgment, unique finger impression distinguishing proof, moving item acknowledgment etc.), e-human services, digital security etc. The significance of digital security is expanding drastically considering the wonderful utilization of online applications, informal organization, Internet of Things (IoT)-based frameworks, and cloud and web innovations. Cyber security is a collection of advancements that secure the between associated frameworks over the web from digital assaults (unapproved individual used to get to information structure online server farms). Digital security is used to forestall harm and assault on cyberspace [virtual personal computer (PC) world that is comprised of many

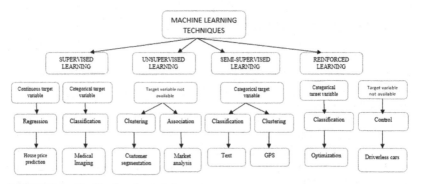

FIGURE 10.1 Machine learning techniques.

system gadgets associated over the web to perform information trade exercises]. The key trait of the internet is an insightful condition for a wide range of members.

The target of cyber security is not only to protect the system from 100% attacks, which is beyond the realm of imagination, but to limit the "attack." The quantity of attackers will be consistently greater than the quantity of individuals attempting to prevent the attacks. Cyber machine learning arrangements must be equipped for tending to well-checked issues. It ought to coordinate with existing apparatuses and architecture. The framework should permit frictionless performance evaluation. Machine learning is getting more noteworthy in the field of digital security. The AI task like relapse can be applied in digital security for misrepresentation identification, grouping strategy can be applied to spam channels, bunching can be applied for legal examination, dimensionality decrease can be utilized for face acknowledgment, and so on.

For the most part, CPS was presented in 2006. CPSs are input frameworks that are smart, ongoing, versatile, or prescient in nature and can be arranged or appropriated. CPS requires improved structure apparatuses that empower plan approach. These systems must help versatility, reduce the intricacy of the executives and furthermore emphasize on check and approval. CPS has applications in various fields like social insurance, apply autonomy, fabricating, transportation, and so forth. Fig. 10.2 speaks about

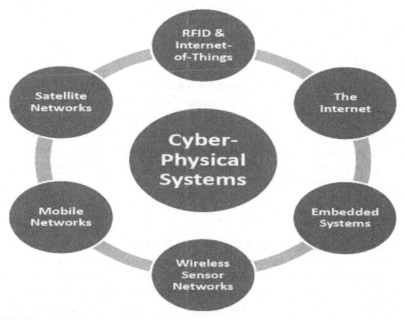

FIGURE 10.2 Overview of Cyber Physical Systems.

various segments of CPS. The principle attributes (Huang, 2008; Krogh, 2008; Li et al., 2009; Rajkumar, 2007) of CPS zones are as follows:

Integrated: CPS has integration of both physical and digital plan.

Resource compelled: The product is inserted in every physical segment and the resources like transmission capacity and calculation speed is limited.

Input controlled: This framework supports high level of mechanization and person − gadget interaction.

Distributed and networked CPS incorporate system of wired or remote system, Bluetooth, GSM etc.

Complex: CPS is carefully obliged by granularity of time and spatiality.

Dynamic reconfiguration: CPS framework is versatile in nature.

Reliable: Since CPS is enormous, convoluted framework, unwavering quality, and security is required a lot because of which the CPS are more reliable.

This segment talks about the distinctive machine learning strategies and how they can be used in CPSs. The rest of the chapter is sorted out as follows: the next section talks about the current work done in this field. Section 10.3 talks about inspiration/motivation behind composing this chapter. In Section 10.4, we discuss the significance of machine learning toward CPS in current scenario and future. Section 10.5 examines a few situations for future with expanded technology (using cyber security). Afterward, we clarified the various difficulties and issues in CPS in Section 10.6. In the same section, we also discuss the openings in this field. In the last section, future aspects of machine learning in CPS, which should be possible in this specific territory, are mentioned.

10.2 Related work

Machine learning strategies have an immense caliber to support digital security. Various machine learning techniques have been effectively utilized to address diverse PC security problems. Machine learning techniques can be used to recognize and avoid brute force attack. Anomaly detection can be applied to identify intrusion detection. We can utilize cluster formation techniques to enhance attack detection. A portion of different applications are quickly discussed in the following sections.

10.2.1 Phishing

Phishing is, where target contacts the individual by means of mail or calls and trap recognizable data, banking and charge card information, and passwords. The data are then used to get to important accounts and can bring about wholesale fraud and money-related loss. Researchers looked at changed machine learning procedures like Logistic Regression (LR), Classification and Regression Trees, Bayesian Additive Regression Trees,

Support Vector Machines (SVM), Random Forests, and Neural Networks (NN). Also, the examinations show that LR has the most elevated exactness and generally high review in correlation with different classifiers. Zhuang et al. (2012) used bunching solutions like various leveled grouping and k-mean clusters for phishing recognition and got 85% performance.

10.2.2 Intrusion detection for networks

The fundamental point of Network Intrusion Detection frameworks is to distinguish vindictive system action, which causes secrecy, uprightness, or accessibility infringement of the frameworks in a network. Subbulakshmi et al. (2010) built Alert Classification System with the assistance of NNs and SVM against Distributed Denial of Service attacks. The specialists guaranteed that normal precision of neural system ready arrangement is 83% while for help vector machines, it is 99%. Sedjelmaci and Feham (2011) propose a half breed answer for recognizing interruptions in a WSN. A grouping method is implemented for lessening the measure of data to process and SVMs with misuse detection strategies are employed for detecting network irregularities.

10.2.3 Key stroke elements validation

Key stroke elements help to distinguish the planning data regarding when the key is squeezed and delivered while an individual is utilizing the console. Revett et al. (2007) proposed applying a Probabilistic Neural Network (PNN) for keystroke dynamics. Also, PNN was contrasted with a multilayer perceptron neural system (MLPNN) with back-engendering. This technique when further examined resulted in MLPNN being just an implementation of PNN multiple times.

10.2.4 Breaking human collaboration proofs (CAPTHAs)

Chellapilla and Simard (2005) propose how the human collaboration proofs (or CAPTCHAs) can be broken by using machine learning. The analysts studied the seven different Hot Isostatic Pressings (HIP) and watched the common qualities and shortcomings. The proposed approach is planned for finding the characters (division step) and employing neural system (Simard, Steinkraus, & Platt, 2003) for character recognition. The tests are led with yippee mail squares, register, Ticketmaster, and Google HIPs. It was accounted that the division stage is moderately difficult due to computational expense.

10.2.5 Cryptography

Yu and Cao (2006) executed a quick and productive cryptographic framework dependent on postponed confused Hopfield neural systems. The scientists guarantee that the proposed framework is made sure about because of "the troublesome synchronization of riotous neural systems with time differing delay." Kinzel and Kanter (2002) clarified how two synchronized neural systems can be utilized for a mystery key trade over an open channel. The specialists have asserted that it is computationally infeasible to play out certain assaults.

10.2.6 Spam detection for social networking

Fake records, mass informing, spreading pernicious links are key to social spamming. Lee, Caverlee, and Webb (2010) saw that spammers abuse social frameworks for utilizing phishing assaults, spreading malware, and advancing partner sites. For ensuring social frameworks against those assaults, a social honeypot was created for recognizing spammers in informal organizations like Twitter and Facebook. The proposed arrangement depends on SVM and has a high accuracy and low false positive rate.

As of now a few methods are available to deal with the unpredictability of CPS. They are referenced briefly.

- Process-based approaches.
- Model-based approaches.
- Architecture-based techniques.
- Organizational approaches.

Even however these strategies function admirably for current frameworks, it cannot withstand the next generation CPS. Hence, this section examines work related to cyber physical framework (counting application and existing methods) in detail. Presently, the next section will examine about inspiration/motivation driving this work.

10.3 Motivation

CPSs are of utmost importance and are utilized fundamentally (by governments/enterprises) to take care of many testing issues of current day-to-day life as it is realized that rapid and right dynamic in a big data system is a major task. Regardless of advantages (and improvement) of CPS, there are numerous issues that we face in our life like absence of brilliant diagnostic apparatuses, which influence healthcare systems. Also they cannot deal with enormous measure of information/large information (which is produced by internet connected device). In addition, individuals with skills are insufficient to deal with or track or investigate these gadgets or frameworks. Likewise, many vulnerabilities may happen on these IoTs/web-associated things. Subsequently, the reconciliation of machine learning in CPSs is an important necessity. Utilizing Machine Learning procedures, skillful individuals can plot or track dangers on web (likewise on CPS) in least time (Checkoway, et al., 2011). We begin composing this chapter because this innovation or

CPS in upcoming future can be helpful in creating proficient and hearty arrangement by means of optimal decisions. This work will concentrate on existing patterns, issues or difficulties in the improvement of modern large information handling, CPS utilizing Machine Learning and digital security in the individual zone. Subsequently, this area talks about our fundamental motivation behind writing this chapter on "Machine learning in Cyber Physical Systems." Further, the next section will discuss about the need (or necessity) of machine learning in CPS.

10.4 Importance of cyber security and machine learning

We can utilize security investigation to identify risks better. It is additionally helpful to organize the cautions and signals that help to discover the arrangement of the issue in less time. We can use machine learning procedure to make the digital security more impressive. Some examples of the models are referenced below:

- Cyber security organizations are utilizing information science techniques to process and examine huge arrangement of information, which can be either notable or threat insight information for a long time.
- F-secure have been utilizing machine learning calculations to take care of issues, for example, order, grouping, dimensionality decrease, and relapse.
- Machine learning can likewise be utilized for effective usage of confirmation frameworks, assessing the convention execution, surveying the security of human cooperation verifications, savvy meter information profiling, and so forth.

Cyber security is a critical territory for the utilization of machine learning procedures. Present day cyber security dangers like malware recognition, interruption location, and information spillage cannot be explained by utilizing scientific models alone. ABI research (a company working on transformational technologies) (Mackenzie Gavel) estimates that "Machine Learning in cyber security will support enormous information, insight, and investigation spending to $96 billion by 2021." In medicinal services, to make sure about all frameworks through which a programmer could get to touchy data (Shamila, Vinuthna, & Amit Kumar), the product would be introduced in the bigger system of the social insurance organization. These frameworks are called endpoints on the grounds that the businesses get to the information at this end. For recognizing digital security dangers, we must introduce machine learning model inside the customer medicinal services organization system and allow to investigate exercises of system progressively. Machine learning creates attention to what ordinary organized action looks like and utilize this as reference to decide the likelihood of dubious movement. If client movement looks like going off track i.e. moving a long way from standard framework, at that point it is considered as misrepresentation action.

Dark trace is an example of machine learning seller for anomaly recognition in healthcare organizations.

By utilizing machine learning, the cyber security frameworks can, without much of a stretch, break down the example and can accomplish information to discover counter measures to stay away from the comparable attacks. In basic terms we can tell that machine learning causes cyber security gatherings to forestall dangers adequately and furthermore to react to dynamic assaults continuously. The job of information is basic for the achievement of machine learning in cyber security. Machine learning is tied in with growing new example and investigating it utilizing various calculations. To accomplish this, we require wide arrangement of information that speaks to numerous expected results from various situations. Note that, at this moment not just the amount but the nature of information matters more. The information must be finished and significant. Presently a day's enormous measure of information is created by the system of IoT and different applications. Conventional framework is not used to deal with this diverse, gigantic information however creating an outline for information can deal with this process viably. In this way, we can infer that, to send compelling digital security framework, machine learning strategies are required, and we need total/significant information to accomplish effective machine learning methods. Along these lines, in this section, we discuss the significance of machine learning methods in digital security. We additionally portrayed the job of information in executing machine learning calculations. In the following section, we will see the eventual fate of CPSs in blend with various existing strategies.

10.5 Machine learning for CPS applications

In the following subsection, we discuss some typical CPS applications that use machine learning:

1. **Smart grid**: It is an unpredictable framework ranging from microscale grid to national or worldwide systems including various levels of facility, the board, and innovation. A keen framework is considered as a CPS as it keeps a track and manages the control generation, stacking, and utilizations through various sensors. These sensors accumulate the stream of information that is taken care of by analytic methods and control frameworks to adjust and convey control generation and consumption (Ellen Nakashima, & Mufson, 2008).

   Due to unpredictability and dynamics of power market, and the volatile nature of sustainable power source, it is critical to have good estimation and calculation of market scenario and energy production to effectively gauge the measure of capacity to produce. In addition to this reason, utilizations of investigation to the smart grid also incorporate flaw recognition at foundation, gadget, framework, and application levels (Sedjelmaci & Feham, 2011).

Machine learning is a capable instrument to investigate the data stream and providing outcomes to inform choices and movements.

2. *Intelligent transportation systems*: This system is a propelled application that expects to offer innovative types of assistance identifying with transport and traffic the executives, and empower clients to be better educated and make more secure, more organized, and more astute utilization of transport systems. Intelligent transportation systems acquire critical improvement transportation framework performance, including reduced congestion and increased safety and traveler convenience (Fox News Network, 2014; https://auto.economictimes; InvestmentWatch, 2014).

Intelligent Transport Systems meets the center qualities of CPS. Empowered by Information and Communication Technologies (ICT), components inside the transportation framework, for example, vehicles, streets, traffic signals, and message signs, are getting savvy by installing microchips and sensors in them. Consequently, this permits correspondences with different specialists of the transportation network, and the utilization of cutting edge information investigation and acknowledgment strategies—to the information gained from implanted sensors. Therefore, shrewd transportation systems empower actors in the transportation system to make better-educated decisions, for example, regardless of whether it is picking which course to take; when to travel; whether to mode-move (take mass travel as opposed to driving); how to advance traffic lights; where to manufacture new streets; or how to consider suppliers of transportation administrations responsible for results (https://auto.economictimes; Zhuo Lu et al., 2010).

3. *Smart manufacture*: Sensor based applications, for example, applications for object identification, power and force sensor-based activities, require exactness of object recognition, present estimation. Also, this exactness needs to breeze through the assessment of time and repeatability (i.e., the results ought to be precise).

Manufacturing smart appliances and car fabricating specifically, requires activities that include taking care of user requirements, investigation or gathering regular information so that the process is deemed to be finished in a couple of moments or seconds. Applications, for example, welding, require continuous information processing, analysis and results. To follow the situation of joining plates on ongoing premise and alter the development of weld firearms becomes the reason for exact and precise welding at high speed (Tobias Hoppe and Dittmann, 2011).

10.6   Future for CPS technology

CPSs are progressively received (actualized) in many automobile organizations and sent with new domains. This details that future CPS is to confront an expansion in useful and extra functional prerequisites. Practical prerequisites incorporate prescient upkeep and expanding computerization levels of CPS. Extra-practical necessities incorporate well-being, digital security, and natural supportability. Numerous new CPS applications are cross-space (capacity to get naturally or send data to other security spaces) in nature. For instance, mechanized vehicles executing apply autonomy advancements and utilization of telecom systems for savvy machines, both represent extraordinary chances and difficulties. Advances in innovation permit totally new kinds of mix and correspondence in CPS over

- technological fields, for example, physical, inserted, organized, and data frameworks. For instance, cloud and edge figuring;
- standalone frameworks, for instance insightful vehicle frameworks (mix of vehicles and foundation);
- stages of the existence cycle, by and large creation information accessible for the duration of the existence cycle and empowering redesigns to programming. These are ideas in Development − Operations (DevOps) mix related with consistent programming advancement, reconciliation, and sending with criticism from operational frameworks (Singh et al., 2020).

There is a quick development in CPS by expanding levels of robotization and knowledge including information investigation. These characteristics depend on the man-made reasoning methods, it incorporates machine learning. AI and information examination openings are viewed as distinct advantages, expressed by a report by the National Science and Technology Council in the United States (Networking and Information Technology Research and Development Subcommittee, 2016). Setting mindfulness is significant for new sorts of AI-based CPS, including the capacity to comprehend that elements are right now part of the close condition and to finish up what their destinations are. The development, furthermore, improvement of AI advances in various application spaces is probably going to drive expanding levels of computerization. Future CPS will likewise be accused of progressively troublesome errands in open conditions due to their capacity to explain cultural difficulties and produce income. To restrict customary assembling applications, profoundly trend setting innovations are conveyed and spread all through society (Kumari et al., 2020). For instance, robotized driving vehicles without human mediation on open streets and robot − human joint effort frameworks.

Keen CPS is answerable for dynamic changes in condition (For instance, profoundly factor traffic conditions that changes quickly with changing human practices and CPS infrastructures.) Which for the most part, implies that not all working conditions are known from the beginning (at the hour of framework advancement) (see model Törngren & Sellgren, 2019). The more extensive suggestions demonstrate that open CPS faces various existing and new types of vulnerabilities from somewhat obscure situations, weaknesses in well-being and assaults (foresee assault by demonstrating "aggressor models"), and altering CPS itself (because of incomplete failures). Uncertainty can apply to perspectives for parts of a CPS and all life cycle stages. By considering different wellsprings of vulnerability, it is essential to design an efficient treatment (Zhang, Selic, Ali, Yue, Okariz, & Norgren, 2016; Molina-Markham et al., 2010).

There is a possibility that the potential attributes of future CPS would require better approaches for thinking with respect to framework level properties and composability. In the following section we discuss multidimensional CPS composability point and portray what we see as explicit composability challenges like human-CPS, CPS incorporating machine learning, reliability, also, Cyber Physical system of systems (CPSoS).

10.6.1   Cyber physical systems and human

Despite the framework type and mechanization level, people are indicating more enthusiasm to cooperate with CPS as designers, administrators, clients, and maintainers. Significant number of difficulties for human − machine connection are brought about by expanding levels of computerization. For instance, now and again people may need to act in crisis circumstances (pilots in airplane). We are as of now moving to more significant levels and more insightful frameworks, and the exercises learned in robotization history remain critical in improving help for human connection with profoundly able CPS (Bainbridge, 1983). A crucial thought for the human CPS is to consider, that is, what decides a specialist's activity or inaction. Deviations in the ordinary conduct of human operators should be recognized, specifically conduct prompts choice/activities for CPS working (Waymo, 2018; Haque et al., 2014).

10.6.2   CPS and artificial intelligence

New kinds of implementations are empowered by Artificial Intelligence and machine learning advancements. While utilizing Artificial Intelligence, the use of neural systems raises question about how to manage heartiness, transparency, predictability, and how to check the effective cost, to approve and guarantee such frameworks (Amodei et al., 2016; Wagner & Koopman, 2015).

10.6.3   Trustworthy

To current CPS, security hazards as of now exist and will increase as CPS is progressively embraced, associated, and utilized by open source software. Absolute security and well-being are unrealistic, so online activities are important to manage security breaks and well-being related anomalies. In expansion, the expanding utilization of CPS makes their accessible progressively significant, which implies that customary hazardous arrangements are not a choice. Future frameworks must execute shortcoming open-minded frameworks while adjusting the ascent for unpredictability because of repetition, flexibility, and fall back measures. At long last, basic issues identified with morals, risk, and confirmation are perceived. Who will assume liability if CPS comes up short, what are the choices made by profoundly robotized frameworks and what is required for future CPS for confirmation case (Waymo, 2018).

10.6.4 Cyber physical systems of systems

Future CPS could be a piece of CPSoS. Such frameworks can likewise be according to various spaces because of their oddity and scale and more prominent number of partner's locales, guidelines, and norms to be mulled over (Engells, 2015).

Subsequently in this section we quickly depict the future extent of cyber physical system in the current period. In the following section we will examine a portion of the difficulties and issues identified with cyber security and furthermore the open doors in detail.

10.7 Challenges and opportunities in CPS

There are numerous issues and difficulties (additionally different chances) for brilliant eventual fate of CPSs in our condition, not many of them will be talked about below:

1. *Security*: Security is a significant test in CPSs for present age and future. All highlights of CPS are associated through web and this is the genuine explanation behind security dangers, that is, everything is all around associated. There is no ideal answer for security assaults until we remain comprehensively associated. If we need to forestall assaults, do not store data in databases that are all inclusively associated, which is unimaginable. In this way, if we remain associated comprehensively, we need to put into security improvisation.
2. *Safety*: CPSs are additionally a potential danger since they are incorporated firmly into the physical condition. Any error can reach out into the physical pieces of the data handling and in some portion of the framework. This can represent a hazard to people, and their data. For instance, stream planes, self-driving vehicles, and gadgets in well-being part. There might be diverse security dangers: they may result from plan mistakes, inappropriate particulars, disappointment of equipment gadgets. To accomplish security, there ought to be some conventional confirmation strategies or utilization of extraordinary dialects for programming like simultaneous dialects (Potop-Butucaru et al., 2006).
3. *Dependency*: Dependability suggests the likelihood that the normal help will be given on schedule. Unwavering quality rules for endured

disappointment levels can appear as roofs. Principle point of trustworthy frameworks is to maintain a strategic distance from administration disappointments, for example, equipment mistakes, natural conditions indicated out of the range to accomplish unwavering quality.

4. *Efficiency*: Significant CPSs are mobiles or the frameworks that are having constrained vitality. Subsequently, such sort of frameworks either have vitality gathering or they need to utilize batteries. In the two cases, usage of vitality ought to be accomplished more cautiously.

5. *Heterogeneous*: CPS connects many devices and different types of components, such as analog components and digital components. Components are designed for different purposes by different companies. Achieving interoperability between components, that is heterogeneity, is a challenge in CPS.

6. *Predictability*: One of the most significant factors in cyber physical systems is time. Also, the planning and conduct of data handling turns out to be basic by connecting data preparing to physical frameworks. Truth be told, most CPS gadgets are progressively frameworks. Within a period of time characterized by nature it is suitable to implement such frameworks. The significance of this prerequisite is regularly ignored. The majority of the frameworks are intended to give snappy reaction time as a rule, however, it may fail because of some other factors. To keep away from such cases for CPS, equipment programming stacks are changed, whole structure may change.

7. *Dynamic in nature*: Work area frameworks are fixed at specific areas. Changing of their system interface occurs in uncommon cases. For CPS, the system association is continuously changing inevitably. Remote LAN, Bluetooth, and mobiles can be utilized by various systems. Each will differ in its speed and cost. The nature of association fluctuates without fail. System deferrals might be extraordinary. Along these lines, arrange association needs to give a significant level of deficiency resilience. The accessibility of vitality and outside gadgets and computational burden additionally change from time to time.

8. *Multidisciplinary*: Structuring of CPS requires different space information; it incorporates material science and software engineering. For most of the applications information is required from disciplines like medication, mechanical designing, and science. Because of tight scholastic for understudy it is beyond the realm of imagination to expect to consolidate all into one. Subsequently, it is very essential to receive CPS ideas in instructive projects with the end goal that all center information is picked up. The combination of registering and the physical condition gives numerous open doors in digital physical frameworks. In the accompanying, by posting not many of the well-known territories, we might want to show the enormous arrangement of openings.

9. **Smart facilities for home**: As for different measurements, there are numerous open doors for edifying life at home. For instance, security levels, vitality productivity, and solace can be improved better, and we can help old individuals. Target of keen home is surrounding helped living. We can improve security and vitality effectiveness by interfacing different gadgets in a home. One of the exceptional cases is zero-vitality building. Primary point of such structures is to create as much vitality as it devours, on normal at any rate. By utilizing savvy ventilation, sunlight-based cells, control of blinds, vitality effective warming, and lighting we can transform this vision into the real world. For genuine utilization of highlights, utilization of vitality is balanced. For instance, air conditioning temperature ought to be less in void rooms.

10. **Logistics and transport**: Transportation is the most well-known utilization of digital physical frameworks. The well-established truth is that in vehicle businesses no vehicle is sold out except if it has a larger number of highlights than the past models; highlights are given by the Information and Computation Technologies (ICT) parts. It incorporates motor control, security highlights like electronic sustainability programs. In self-governing driving vehicles, the principle include is leaving help (Okuda et al., 2014). We have a ton of transport-related coordinations, gracefully including chain enhancement and without a moment to spare conveyance. Because of expanding number of clients requesting through web, productive bundle conveyance is more significant. For this, ICT parts are bolstered to store and recovery of products (Auffermann, Kamagaev, Nettsträter, tenHompel, Vastag, Verbeek, & Wolf).

11. **Healthcare systems**: The well-being part incorporates countless different utilizations of this coordination. New ICT-empowered sensors can be planned. We can have propelled procedures for information and hazard examination. [In community research focus SFB 876, these are considered (Morik et al., 2015)]. To recognize wellsprings of issues, gracefully chains can be checked. After finding is finished, it is additionally conceivable to help treatments. Everyone (even crippled patient) can have customized drugs. Results can be observed by the utilization of sensors. The point-by-point data of patient are given in understanding data frameworks, which stays away from repetitive data about patient.

12. **Disaster management**: ICT (Information and Computation Technologies) used to give salvage activities to the catastrophes, in which correspondence assumes as a key job. Note that a few issues and difficulties toward CPS and Medical Cyber Physical Framework additionally have been talked about by Tyagi (2016) and Meghna et al. (2019). Thus in this section we secured various difficulties and issues existing in this field. Apart from this we additionally referenced the chances of the digital physical framework. In the next section we will sum up this work.

10.8 Conclusion

With the quick turn of events, we require smart support frameworks to take care of assembling and numerous control applications. As we probably are aware, man-made applications can help lessen human work power. It is also imperative in identifying weaknesses or danger on CPSs. Digital security is important for each business (particularly that which is associated/working through web), since at present pretty much every business is influenced by digital violations. Future brilliant enterprises will require to advance not just their own assembling forms but also the utilization of items and assembling resources. On another side, the combination of CPSs and intelligent product is needful for two true modern cases, which can cover various periods of the item life cycle like the creation, use what is more, and upkeep stages. In rundown in the not so distant future, the incorporation of cutting-edge examination utilizing present day devices can improve assembling or industry 4.0 (into assembling, items, and administrations). This region is still in the developing stage, and thus requires consideration from new researchers and established researchers around the globe. In this way, intrigued individuals are sympathetically welcomed to accomplish their exploration, take a shot at above recorded open issues openings toward the coordination of cyber security, artificial intelligence (or machine learning), and CPS.

References

Abu-Nimeh, S., Nappa, D., Wang, X., & Nair, S. (2007, October 4−5). A comparison of machine learning techniques for phishing detection. *APWG eCrime Researchers Summit*. Pittsburg, PA.

Amodei, D., Olah, C., Steinhardt, J., Christiano, P., Schulman, J. & Mané, D. (2016). Concrete problems in AI safety, arXiv, arXiv:1606.06565.

Auffermann, C., Kamagaev, A., Nettsträter, A., tenHompel, M., Vastag, A., Verbeek, K.& O. Wolf, Cyber physical systems in logistics, http://www.effizienzcluster.de/files/9/5/938_scientific_paper_cyber_physical_systems_in_logistics.pdf.

Bainbridge, L. (1983). Ironies of automation. *Automatica, 19,* 775−779. Available from https://doi.org/10.1016/B978-0-08-029348-6.50026-9.

Checkoway, S., McCoy, D., Kantor, B., Anderson, D., Shacham H., Savage, S., Koscher, K., Czeskis, A., Roesner, F., and Kohno, T. (2011). Comprehensive experimental analyses of automotive attack surfaces. In *USENIX Security Symposium*.

Chellapilla, K., & Simard, P. Y. (2005). Using machine learning to break visual human interaction proofs (HIPs). *Advances in Neural Information Processing Systems, 17,* 265−272.

Ellen Nakashima., & Mufson, S. (2008). Hackers have attacked foreign utilities, cia analyst says. *Washington Post (Washington, D.C.: 1974), 19.*

Engells, S. (2015). European Research Agenda for Cyber-physical Systems of Systems and their Engineering Needs. Available online: http://www.cpsos.eu/wp-content/uploads/2016/06/CPSoS-D3.2-PolicyProposalEuropean-Research-Agenda-for-CPSoS-and-their-engineering-needs.pdf.

Fox News Network. *Threat to the grid? Details emerge of sniper attack on power station* (2014). http://www.foxnews.com/politics/2014/02/06/2013-sniper-attack-on-power-grid-still-concern-inwashington-andfor-utilities/.

Haque, S. A., Aziz, S. M., & Rahman, M. (2014). Review of cyber-physical system in healthcare. *International Journal of Distributed SensorNetworks.*

Hoppe, T., Kiltz, S., & Dittmann, J. (2011). Security threats to automotive can networks—Practical examples and selected short-term countermeasures. In *Proceedings of the 27th international conference on computer safety, reliability, and security, SAFECOMP '08* (pp. 235–248). Berlin, Heidelberg,. SpringerVerlag.

Huang, B.X. (2008, June). Cyber physical systems: A survey. *Presentation report.* https://auto.economictimes.indiatimes.com/news/industry/the-future-of-autonomous-vehicles-in-indi asteering-the-legal-issues/64985989.

InvestmentWatch (2014). First time in history, a terrorist attack on the electric power grid has blacked-out an entire nation in this case yemen. http: //investmentwatchblog.com/first-time-in-history-a-terrorist-attack-onthe -electric - power- grid - has - blacked - out - an - entire - nation - in - this - case-yemen.

Kinzel, W. & Kanter, I. (2002, November 18–22). Neural cryptography. In *Proceedings of the 9th international conference on neural information processing*, Vol. 3 (pp. 1351–1354)

Krogh, B.H. (2008). Cyber physical systems: The need for new models and design paradigms. *Presentation report.* Carnegie Mellon University.

Kumari, R., Kumar, S., Poonia, R. C., Singh, V., Raja, L., Bhatnagar, V., & Agarwal, P. (2020). Analysis and predictions of spread, recovery, and death caused by COVID-19 in India. *Big Data Mining and Analytics, IEEE.* Available from https://doi.org/10.26599/BDMA0.2020.9020013.

Lee, K., Caverlee, J., & Webb, S. (2010, July 19–23). Uncovering social spammers: Social honeypots + machine learning SIGIR'10. Geneva, Switzerland.

Li, J. Z., Gao, H., & Yu, B. (2009). Concepts, features, challenges, and research progresses of CPSs. *Development Report of China Computer Science*, 1–17.

Lu Z, Lu X., Weng W., & Weng W. (2010). Review and evaluation of security threats on the communication networks in the smart grid. In *Military Communications Conference—MILCOM 2010* (pp. 1830–1835). IEEE.

Mackenzie Gavel, ABI research blog, https://www.prnewswire.com/.

Meghna, M.N., Tyagi, A.K., Richa, G. (2019). Medical cyber physical systems and its issues. International conference on recent trends in advanced computing, ICRTAC 2019, Procedia Computer Science 00 (2019) 000-000, 2019.

Molina-Markham, A., Shenoy, P., Fu, K., Cecchet, E., & Irwin, D. (2010). Private memoirs of a smart meter. In *Proceedings of the 2nd ACM Workshop on Embedded Sensing Systems for Energy Efficiency in Building* (pp. 61–66). ACM.

Morik, K. et al., Collaborative research centre on resource constrained machine learning, 2015, http://www.sfb876.tu-dortmund.de.

Networking and Information Technology Research and Development Subcommittee (2016). The National Artificial Intelligence Research and Development Strategic Plan; Executive Office of the President or the United States: Washington, DC, USA.

Okuda, R., Kajiwara, Y. & Terashima, K. (2014). A survey of technical trend of ADAS and autonomous driving. In *Proceedings of technical program, 2014, International Symposium on VLSI Technology, Systems and Application (VLSI-TSA)* (pp. 1–4). doi:10.1109/ VLSI-TSA.2014.6839646.

Potop-Butucaru, D., de Simone, R., & Talpin, J. P. (2006). The synchronous hypothesis and synchronous languages. In Z. Richard (Ed.), *Embedded systems handbook.* CRC Press.

Rajkumar, R. (2007, May). *CPS briefing*. Carnegie Mellon University.

Revett, K., et al. (2007). A machine learning approach to keystroke dynamics-based user authentication. *International Journal of Electronic Security and Digital Forensics*, *1*(1).

Sedjelmaci, H., & Feham, M. (July 2011). Novel hybrid intrusion detection system for clustered wireless sensor network. *International Journal of Network Security & Its Applications (IJNSA)*, *3*(4).

Shamila, M., Vinuthna, K. & Amit Kumar, T. A review on several critical issues and challenges in IoT based e-healthcare system. *IEEE*.

Simard, P.Y., P.Y. Steinkraus D. & J. Platt. (2003). Best practice for convolutional neural networks applied to visual document analysis. In *International Conference on Document Analysis and Recognition (ICDAR)* (pp. 958−962). IEEE Computer Society, Los Alamitos.

Singh, V., Poonia, R. C., Kumar, S., Dass, P., Agarwal, P., Bhatnagar, V., & Raja, L. (2020). Prediction of COVID-19 coronavirus pandemic based on time series data using support vector machine. *Journal of Discrete Mathematical Sciences & Cryptography*. Available from https://doi.org/10.1080/09720529.2020.1784525.

Subbulakshmi, T., Shalinie, S. M., & Ramamoorthi, A. (2010). Detection and classification of DDoS attacks using machine learning algorithms. *European Journal of Scientific Research*, *47*(3), 334−346, ISSN 1450216X.

Törngren, M. & Sellgren, U. (2019). Complexity challenges in development of cyber-physical systems, In Principles of Modelling, Springer: Cham, Switzerland, 2018, Vol. 10760, doi:10.1007/978-3-319-95246-8_27. Computer Reviews Journal Vol 4, ISSN: 2581−6640 http://purkh.com/index.php/tocomp 76.

Tyagi, A. K. (2016). Cyber physical systems (CPSS)—Opportunities and challenges for improving cyber security. *International Journal of Computer Applications*, *137*(14).

Wagner, M., & Koopman, P. (2015). *A philosophy for developing trust in self-driving cars. In Road Vehicle Automation2* (pp. 163−171). Cham, Switzerland: Springer. Available from 10.1007/978-3-319-19078-5_14.

Waymo, W. (2018). Safety report: On the road to fully self-driving. Available online: https://storage.googleapis.com/sdcrod/v1/safetyreport/Safety%20Report%202018.pdf.

Yu, W., & Cao, J. (2006). Cryptography based on delayed chaotic neural networks. *Physics Letters. A*, *356*(4−5), 333−338, ISSN 0375−9601.

Zhang, M., Selic, B., Ali, S., Yue, T., Okariz, O., and Norgren, R. (2016). Understanding uncertainty in cyber-physical systems: A conceptual model. In *Proceedings of the 12th European Conference on Modelling Foundations and Applications*, Vienna, Austria, July 4−8, 2016, Springer-Verlag: Berlin/Heidelberg, Germany, Vol. 9764 (pp. 247−264), doi:10.1007/978-3-319-42061-5_16.

Zhuang, W., Ye, Y., Chen, Y., and Li, T. (2012, December 17). Ensemble clustering for internet security applications. In *IEEE Xplore*.

Chapter 11

Impact analysis of COVID-19 news headlines on global economy

Ananya Malik, Yash Tejas Javeri, Manav Shah and
Ramchandra Mangrulkar
Computer Engineering Department, Dwarkadas J. Sanghvi College of Engineering, Mumbai University, Mumbai, India

11.1 Introduction

The novel coronavirus or COVID-19 is a global pandemic that has affected a large number of the world population. Apart from its ravaging rate of infection, the pandemic has changed the way the world functions. Starting from early March 2020, almost 161 countries have gone under lockdown (Elliott, 2019). 90% of the world's schools suffered because of the lockdown (Strauss, 2020). The lockdown was variable in each country, with countries like Italy, the United Kingdom, and India entering a state of complete lockdown. Despite the lockdown being complete or partial, the impact this period has had on the economy is not fully comprehensible. With businesses shut down, approximately 140 crore people lost their jobs across all sectors, and those who were able to, tried working from home (International Labour Organisation, 2020). The International Monetary Fund compared the economic depression caused by the lockdown to that of the recession of 2008 and the great depression (Georgieva, 2020). In 2008, the fall in global growth was projected to be −0.1% as compared to the 2020s at −3.0%. Although the economy around the world has been heavily impacted due to the coronavirus pandemic, the exact comprehension of the extent of this impact is not known.

Efficient market hypothesis suggests that the share prices reflect all information. Observations have been made throughout decades correlating the impact of news articles on the stock market. In April 2013, hackers hacked the United States' Associated Press's Twitter account and tweeted the headline "Two Explosions in White House and Barack Obama is injured" (BBC News, 2013).

Cyber-Physical Systems. DOI: https://doi.org/10.1016/B978-0-12-824557-6.00001-7
189

This caused the DJIA index to drop 100 points within seconds of the release. The above example is proof to show how headlines and news analytics have had a strong influence on affecting the stock market. News articles have had a stronger effect on the overall market performance; as compared to tweets, news articles are considered more trustworthy and permanent. Economic growth and the stock market are seen to be correlated. A long period of good economic growth is reflected upon the increasing stock price, and falling shares indicate a slowing economy. The Natural Language Processing (NLP) community has been exploring this area of research for a while trying to produce a state-of-the-art algorithms and systems that can be used to correlate as well as predict the stock directly from the news articles of the corresponding day. This chapter aims to contribute to the research topic of the correlation of the news articles to the economy, with a focus on the current situation of the coronavirus pandemic. An important reason for working on this topic is that the preexisting systems of analyzing sentiments on text fail for the headlines focused on the coronavirus pandemic. The news headlines surrounding the current pandemic contain medical jargons that are not correctly interpreted by the state-of-the-art sentiment analysis models. For example, the phrase "tested positive" contributes toward a positive sentiment value; however, in the context of the pandemic, it should contribute toward a negative sentiment value. This chapter aims to present an efficient solution to solve this issue. The chapter contributes toward creating a framework to generate and predict stock indices. Unlike other works in the area, this model focuses on generating corresponding stock indices with a greater focus on the current coronavirus medical crisis.

This chapter consists of the subsequent sections such as Section 11.2 focuses on exploring the related work in this field. Section 11.3 highlights the methodology adopted in this chapter. Section 11.4 is dedicated to validating and understanding the results of the system. This chapter is concluded by discussing the challenges as well as paving the path for future scope in Section 11.5.

11.2 Related work

Joshi, Bharathi, and Rao (2016) explored the efficient market hypothesis to correlate news articles to the subsequent rise and fall in stock prices. This chapter uses fundamental analysis for prediction, that is depends upon news articles as well as the daily open and close price of the stocks of the given company collected over a long period of time. The entire mechanism is divided into three phases. In the first phase, they propose preprocessing of the news data of the "Apple Inc Company" to convert the raw, unstructured text, into readable and labeled text. In the next step, using a sentiment detection algorithm, the sentiment of the article is detected. This step is to assign a polarity score to the respective articles. The above output serves as an input for the second phase in which the preprocessed data are converted into a document vector. The text vector is then trained on three classification algorithms: Naive Bayes, random forest, and support vector machine (SVM). This is then tested on sets of data of different time periods to compare the accuracy and other metrics. To provide a comparison, the predicted news score is compared with the historical values of the stock. The above system, when used with the random forest algorithm, provides the best accuracy at 88%−92% and can successfully predict the stock trends.

Im, San, On, Alfred, and Anthony (2014) aim to highlight a lexicon-based approach for sentiment analysis. The chapter proposes manually creating a lexicon and adding words that match the intended domain; in this case, this domain lies around the financial and stock market. The lexicon contains the frequently tagged words that are obtained from multiple articles. For experimental purposes, they used three different kinds of datasets: content only, headline only, and content and headline only. With this, they obtained 8000 tagged words. This lexicon then undergoes data preprocessing to remove noises such as stop words, redundant words, etc. To normalize the system, "Porter Stemmer" algorithm is used to eliminate any morphological endings. Each word in the lexicon is tagged with a positive or negative polarity based on matching with the pretagged words. Finally, for the given article, the overall sentiment is calculated using the positive or negative sentiment ratio. This method provides an accuracy score of 79.1% against the actual market values. One disadvantage of using a lexicon-based approach is that it only considers one word at a time; thus it does not consider the possibility of a previous word affecting the sentiment, which is not accurate.

Kirange and Deshmukh (2016) propose to use emotion recognition to predict the stock price of the market. The dataset consists of articles that are accumulated from various resources and news sites. An emotion dictionary is maintained to classify the words as positive or negative. The emotion dictionary used here is the dictionary of Linguistic and Inquiry Word Count (LIWC). This dictionary is used to create the feature matrix, which, along with the human emotion annotation, is used to train a classifier. The classifier is now allowed to trim the dataset offline. This system produces positive and negative emotion and is tested on various classifiers like K-Nearest Neighbors (KNN), Naive Bayes, and SVM. The predictions used from the above classifications on the sentiment score is used to predict the stock price for the next day. This graph curve is tested with historical data of 10 years.

For positive news, the highest accuracy achieved is 62.72%, and for negative news, the highest accuracy achieved is 69.73%. A limitation of the above method is that while the accuracy is satisfactory, an overall market view can be taken into consideration by incorporating social and legal parameters.

To analyze the effect of news sentiments on the stock market, Shah, Isah, and Zulkernine (2018) use a methodology that involves a dictionary-based approach. The system proposed by them involves five steps in which they analyze how to process data based on pharmaceutical companies. Data are collected via web scraping from different financial websites. These data are then processed to clean it and tokenized to suitable formats. This is then compared with a dictionary designed especially for medical phrases and words, thus analyzing the sentiment and score. The score is then identified and plotted against the stock data. The model finally is able to predict the result as a classification of the decision to be taken into three categories, buying the stock, selling it, or retaining it. The above system, when used gives 70.59% accuracy in predicting the daily directional (up/down/neutral) changes in the stocks.

Souma, Vodenska, and Aoyama (2019) take into consideration the entire articles and not only the headlines. They choose to compare articles with the stock values of two major indices like Thomson Reuters News Archive and Thomson Reuters Tick History. Their approach included the following steps of creating a word embedding, defining the polarity, preparing the training and testing data, and finally applying the deep learning model. They choose their data from the Wikipedia 2014 and Gigaword 5 corpus and have defined the polarity of the articles directly based on the stock prices. They set the final max length of the training data at 50 words. Finally, for the prediction, they applied Long Short-term Memory network units to predict the stock prices. They selected the training data either randomly or hierarchically, in which the hierarchical performance performed better with a 97.5% accuracy as compared to the random approach with an accuracy of around 95%.

Kalra and Prasad (2019) establish a correlation between circulating news and fluctuations of price in several stocks in the financial industry. They propose a method that analyzes sentiments from news articles along with their targeted stock's change in closing price. Extrapolating from historical data, they predict the price of a stock using a supervised model. First, they calculate negative and positive sentiments from news articles for a particular stock for a particular day, which they then feed to a machine learning model along with stock prices to comprehend the influence the news article had by observing price changes. They have compared three such models, namely KNN, SVM, and Neural Network. The KNN model provided an accuracy of 75% − 91.2% for varying stocks. Similarly, the SVM model's accuracy was in the range of 65% − 83.80%, and the Neural Network's accuracy was between 73.80% and 88.70%. The paper claimed that variance had a substantial impact on the movement of stock prices. The paper showed a predictable trend in stock prices based on news articles, which could be of monetary benefit.

Khedr, Salama, and Yaseen (2017) aim to help investors maximize profit by predicting the best time to purchase and offload shares. By analyzing text from news articles, they are able to determine the "polarity" of news pertaining to a specific stock. They have also taken into consideration historical data such as the highest and lowest value during a day along with the opening and closing price. The authors used three NASDAQ stocks: MSFT, FB Inc, and Yahoo Inc for their analysis. They collected news for sentiment analysis from sources like The Wall Street Journal and Reuters for a total of three pieces each day. After preprocessing text, applying N-gram feature extraction, and finding the weighted value of each term, a naive Bayes classifier finally determines if the text is positive or negative in context. Their reasoning behind using this model is its swiftness. Finally, a KNN machine learning model was trained to predict the trend of the stock, whether the stock was about to rise or fall and what the expected variation in price was. Using this model, they were able to achieve an accuracy of 86.21%, 82.76%, and 72.72% for Yahoo, MSFT, and FB Inc, respectively.

Li, Xie, Chen, Wang, and Deng (2014) aim to speculate stock prices based on sentiment analysis of articles published in the media. They then outline how they try to reduce dimensionality by making comparisons between thousands of words. In an experiment to find out a suitable approach, they tested six variations in total, consisting of two different dictionary types: one using SenticNet and the other using bag of words. Further, they showcase two traditional polarity-based methods as well. They have based their analysis on open-to-close prices of a share instead of close-to-close prices. A radial basis function kernel SVM is trained on this historical data and its parameters, gamma, and C are tuned using the grid-search method. A Gaussian curve is assumed to determine a reliable threshold. Overall, they found that the Harvard IV (HVD) sentiment dictionary provided better results than the bag of words approach, on an individual stock comparison. However, on a sector level and index-level basis, Laughren–McDonald sentiment dictionary was found to be better suited than HVD. Finally, they were able to show that sentiment analysis helped predict stock price fluctuations more accurately.

Mankar, Hotchandani, Madhwani, Chidrawar, and Lifna (2018) propose the prediction of stock market values using analysis of social media text. For this method, the researchers suggest an analysis of tweets that are collected using the Twitter API and stored in the JavaScript Object Notation format. The tweets are processed, and keywords have been extracted using spaces and stop words. These preprocessed data are then pushed into the next phase of computation to obtain valuable features and create a feature matrix. The paper first tries classification on Naive Bayes and SVM algorithms to create the feature set. This system can classify the word as positive or negative, depending upon the feature set. One of the biggest problems the researchers faced here is that with increasing data, handling the data was getting more challenging. Thus an appropriate solution proposed is to use a python library called AFINN for training dataset labels, to classify into positive or negative. Therefore now, each tweet is assigned a class, and the total for the day is calculated. The researchers then use Nifty data to correlate with the sentiment values. While the system is able to provide an accurate relation between the index and sentiment, due to a lack of historical and real-time tweet data, it is unable to predict accurate values.

Bhat et al. (2020) have analyzed the sentiments of social media during the time of the coronavirus pandemic. They have used tweets as their primary source of information to analyze the mood since the first few global cases of the virus emerged in February 2020. They retrieved tweets with the primary two hashtags of COVID-19 and coronavirus and analyzed them in the given period. The analysis revealed that close to 51% of the tweets were positive, 31% neutral, and 14% negative. From this result, an inference could be made with the understanding that despite the lockdown and subsequent uncertainty that exist, people are hopeful of the world returning to normalcy and the positivity is also addressed as an appreciation toward the efforts of the medical community. This chapter sets the path for future work to extrapolate and make decisions in accordance with social media sentiments.

11.3 Proposed methodology

The proposed methodology is divided into six parts and is shown in Fig. 11.1. The first step is the collection of news articles from sources. These data must consist of article headlines and date of publishing. Before the analysis is performed, these data must be preprocessed. Then, the sentiment of each article, using sentiment analysis algorithms, is analyzed and then grouped based on the date of publishing. This is done to estimate the sentiment score for each given day by taking the average of the sentiment scores of all the headlines for that given day. This chapter then proposes combining the averaged sentiment scores with the Nifty index's closing values for each day over the last 4 years. Various regression algorithms are then applied in an attempt to predict the Nifty index for future headlines. In order to validate the proposed method, this chapter presents a case study on the Indian market and economic conditions and has used the Nifty index as a valid indicator.

11.3.1 Data and data preprocessing

This work aims to portray the impact of COVID-19 on the global economy; hence the data consist of the news headlines from those months in which the virus caused the world to undergo a lockdown. Data from April 2020 to June 2020 has been considered for this study. The data source (Prasad, 2020) by Jayanti Prasad contains news headlines from various English newspaper publications such as Times of India, Business Standard, The Hindu, etc. It has 2575 instances of news headlines focused on the pandemic, out of which 2060 and 515 entries are used for training and evaluating the model, respectively. The data are then cleaned and preprocessed. An essential part of preprocessing is to make sure the dataset lies within the conformity of the dictionary of the pretrained model. The obtained values are for the historical Nifty index from Yahoo Finance. The area of

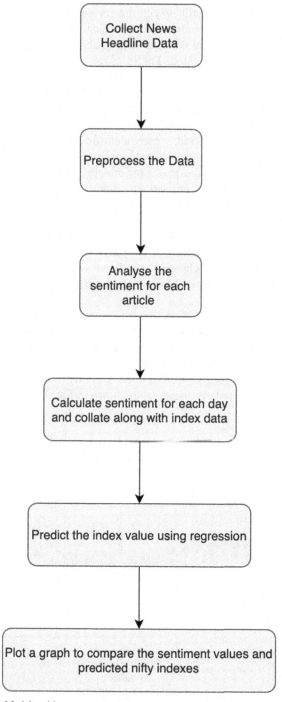

FIGURE 11.1   Model architecture.

focus of this chapter is headlines with the coronavirus news. While analyzing the dataset, the work finds that the medical terminology often contradicts the dictionary used by the sentiment analysis algorithm used. Hence, the dataset has been reoriented. For instance, in "12 new positive cases found," "new" and "positive" are both regarded as good sentiments, and thus a positive sentiment would be generated. However, in the context of coronavirus, the above statement should correspond to a negative sentiment value. Thus to avoid such problems, a word cloud has been created of the headlines in a 4-month period, which identifies the most frequently used words and changes the context accordingly. In addition, to compare sentiments during and before the pandemic months, a comprehensive dataset (Rohk, 2017) by Rohk was utilized. It contains 121,294 Indian news articles leading up to the pandemic, consisting of fields such as the date of publication and the news headline.

These two datasets can be visualized in the form of the following graphs: Fig. 11.2 represents the change in sentiment leading up to the pandemic while Fig. 11.3 depicts the change in sentiment with time during the months of the pandemic, and Fig. 11.4 is the graph for change in the Nifty index with time.

In the graph in Fig. 11.2, there seems to be a clear downward inflexion in the mood of the nation when cases began rising and the lockdown was implemented, as perceived by analyzing newspaper headlines. Although sentiments have occasionally and slightly improved during the lockdown, such as when workplaces were allowed to open, sentiments during these challenging times have majorly been negative, as shown in Fig. 11.3.

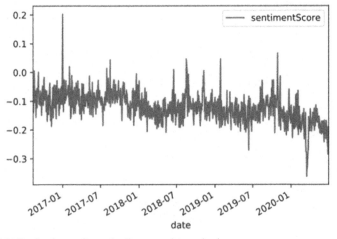

FIGURE 11.2 Sentiment change leading up to the pandemic.

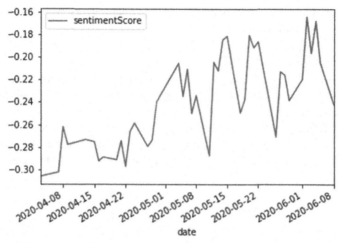

FIGURE 11.3 Sentiment change with time during the pandemic.

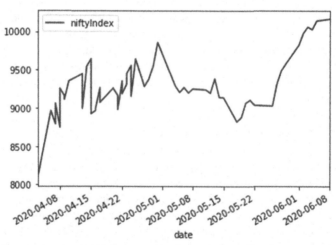

FIGURE 11.4 Nifty index change with time during the pandemic.

11.3.2 Sentiment analysis

An essential aspect of this task is to be able to evaluate the meaning and impact of the news headlines. In order to analyze this, this chapter proposes the analysis of the sentiment of the headline, that is. whether the given sentence is positive, negative, or neutral. Unlike other works, this task aims to predict the final index value. Hence the research proposes to calculate the final sentiment score of the text instead of focusing on just the class of sentiment. There are three methods to analyze sentiments. Machine learning, deep learning, and lexicon analyzer.

11.3.2.1 Machine learning

In this method, sentences are represented in the form of vectors, which represent both the words as well as the semantic structure of the sentence. This can be obtained using Doc2Vec, an NLP tool that represents the text data in a vector form. The semantic features can then be accessed from this vector representation. Using these semantic features of the sentence, the sentiment classes can be obtained, that is very negative, negative, neutral, positive, or very positive.

11.3.2.2 Deep learning

Although sentences are represented in the form of vectors in the deep learning approach as well, this model processes the words from the first word to the last for the sentiment. Each vector contains the memory, that is remembering the words that came before as well as the partial output. "Long Short-term Memory Recurrent Neural Networks" is one such example. This accounts for the contexts in sentiment analysis. While the accuracy of this system is usually good, one of the disadvantages of using such a system is that it needs labeled or explicitly tagged data for training, corresponding to each news headline in the dataset. The headline dataset that this chapter makes use of directly retrieves data from news websites, and hence, its labels are not available.

11.3.2.3 Lexicon method

Lexicons are often looked at as a rule-based approach. The basic procedure followed by the lexicon-based process involves the tokenizing of the data. The words in the sentence are converted into tokens, and the stop words and the punctuations are removed. A sentiment lexicon is maintained, which holds the sentiment values of each word and can judge based on objectivity and intensity as well. For the given application, this chapter uses the lexicon-based approach for calculating the sentiments over the other two methods, because of the availability of the lexicons and its fast processing. Moreover, it does not require a labeled set of data. This work has tested sentiment analysis for the given COVID data with two lexicon-based approaches "TextBlob" and "VADER." To calculate the sentiment score, three main parameters are considered: polarity, intensity, and subjectivity.

Polarity is a measure of whether the given sentence is positive or negative. Intensity or valency is a measure of how positive or negative the sentence is.

Subjectivity is a measure of how opinionated is the context of the sentence. TextBlob is a sentiment analysis library that comes along with the Natural Language Toolkit (NLTK) library. TextBlob returns the polarity on a scale of -1.0 to 1.0 and subjectivity on a scale of $0.0 - 1.0$. For the use case, the polarity and intensity were of utmost importance since accurate scores are needed in predicting the stock sentiment. Thus due to restrictions in the range, the TextBlob algorithm is not used. Valence Aware Dictionary and Sentiment Reader (VADER) is an extremely popular algorithm of the NLTK library. One of the main advantages of VADER is that it is constructed for use in social media. Social media text is usually categorized as a short form text, which should align with news headlines. VADER is sensitive to polarity and intensity and measures it in the range of -4.0 to 4.0 from negative to positive. Thus out of the two lexicon-based approaches, the chapter makes use of VADER for the given application.

Thus post the sentiment analysis phase, the output is a dataset of the news headline and the sentiment score calculated for each news item. The final aim is to predict the indices from the news article. Hence to achieve the final dataset, the sentiment score for each day is averaged. All the articles for the given day are taken and sum the actual sentiment score for each day grouped according to published date and divide it with the number of articles. This ensures that both positive and negative articles are given equal fitting. This dataset is combined with the closing values of the index for the corresponding date. The values for the historical Nifty index are obtained from Yahoo Finance. Therefore the final data consists of columns, such as the Date, Average sentiment score, and the Nifty index value.

11.3.3 Prediction of Nifty score

The Nifty index is frequently used to assess the strength of the Indian economy as it is a metric that tracks and takes into account important sections of the industry that impact the nation's economy. Therefore in order to fully comprehend the impact of coronavirus on the economy, this chapter tries to predict the Nifty index value on a given test data. This would help establish how declining sentiments caused by the coronavirus pandemic is affecting the Nifty index. This, in turn, would indicate how the economy is being affected on a larger scale. For this purpose regression, a statistical analysis method to estimate the relationship between a dependent variable and one or more independent variables is used. "Closing value of Nifty index" for a particular day or "change in the Nifty index with respect to the previous day," either can be considered for training. By experimentation, it is observed that taking the closing Nifty index values performed better than taking the net change in the Nifty index because the latter only shows by how much it has changed, irrespective of the range at which Nifty is present during that particular period. Hence, the dependent variable here is the Nifty index, and sentiment score is the independent variable, which is also illustrated in Fig. 11.5. With regression, the aim is to predict the exact index for each day. To test the system, the model in this chapter is trained using linear

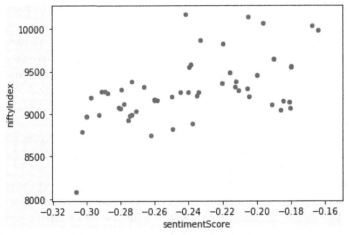

FIGURE 11.5 Nifty index versus sentiment score.

regression, random forest, polynomial regression with degree 3, and the gradient boosting algorithm.

11.3.3.1 Linear regression

Linear regression is a model that attempts to describe the linear relationship between the independent and dependent variables. This is a cause and effect relationship between the features and the target variable. A linear regression line has an equation of the form $Y = mX + c$, where X is the independent variable and Y is the dependent variable. The slope of the line is m and c is the intercept. For instance, linear regression can be used to relate the height and weight of individuals linearly.

11.3.3.2 Polynomial regression

Linear regression is used to establish linear regression between the variables. However, if the distribution of the data is nonlinear, polynomial regression can be used. It is an extension of general linear regression that aims at looking at the relationship between the dependent and the independent variable, modeling it at the nth degree polynomial. A polynomial regression curve has an equation of the form $Y = m_0 + m_1X + m_2X^2 + m_3X^3 + \ldots\ldots + m_nX^n + c$, where X is the independent variable and Y is the dependent variable. All the ms are the coefficients of the respective degrees of the independent variables, and c is the intercept.

11.3.3.3 Random forest

Random forest regression is a state-of-the-art machine learning technique that can be used for both the classification and regression. Forest is made up of trees and random forest builds multiple decision trees and merges them together to get a more accurate and stable prediction (a decision tree is a flow-chart used to visually and explicitly represent decisions and decision making). It is an ensemble method that is better than a single decision tree because it reduces the overfitting by averaging the result. It uses a statistical method called bagging, which reduces the complexity of the model, preventing overfitting of the data. The general idea behind bagging is that a combination of learning models increases the overall result, that is random forest averages the predictions of multiple trees allowing information sources to be expanded.

11.3.3.4 Gradient boost regressor

Gradient boosting is a machine learning technique for regression and classification problems, which like random forest also predicts using outputs from multiple decision trees. Gradient tree uses the decision trees as weak learners. Gradient boosting utilizes the concept of residuals to calculate the difference between the current prediction and known target value. After calculating the residual, the algorithm will map the weak features to the residual and pushes the model toward the target value, as this step is repeated multiple times. Unlike random forest, gradient boost regressor creates one tree at a time and corrects the errors of the previous three. While it is computationally more expensive, the final output is more accurate. This method is used in this chapter as it is used to predict continuous values like heartbeats and stock indices.

11.4 Results and experimental framework

This work uses Python as the main programming language for implementing this concept. Various libraries have been used. NumPy and Pandas are both used to facilitate data manipulation and easier access to data, Matplotlib and Seaborn are used for data visualization, and the Sklearn package has been used to import the regression algorithms for prediction. This chapter aims to get the best possible predictions out of each model by tuning their hyperparameters, train-test split, etc.

11.4.1 Linear regression

In implementing a model to predict Nifty, a linear regression model is trained on the above-described dataset. The linear regression model is imported from the sklearn package as "sklearn.linear_model." For linear regression, the dataset is divided into training and test data in the ratio of 80:20. With linear regression, the root mean squared error (RMSE) obtained is 201.785 and R^2 value is 0.648. This shows that there is some correlation between the Nifty index and sentiments. Fig. 11.6 shows the plot between

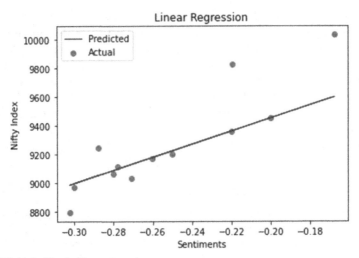

FIGURE 11.6 Plot for linear regression.

the sentiment value and the predicted Nifty index value for the linear regression model.

11.4.2 Polynomial regression with degree 3

With an attempt to improve upon the predictions of the linear regression model, a polynomial regression of degree 3 is trained. The polynomial regression's preprocessor is imported from the sklearn package as "sklearn. preprocessing.PolynomialFeatures" and the dataset is divided into training and test data in the ratio of 80:20. With polynomial regression, the RMSE obtained is 214.548, and R^2 value as 0.602. Fig. 11.7 shows the plot between the sentiment value and the predicted Nifty index value for the polynomial regression model with degree 3.

11.4.3 Random forest regression

Random forest regression is also used to try and improve the accuracy over linear regression as random forest will certainly be able to approximate the shape between the targets and features. The random forest regression model is imported from the sklearn package as "sklearn.ensemble. RandomForestRegressor." By experimenting, it was found that changing the split of training data and test data from 80:20 to 70:30 resulted in the minimum loss. With random forest regression, the RMSE obtained is 299.652 and R^2 value is 0.215. Fig. 11.8 shows the plot between the sentiment value and the predicted Nifty index value for the random forest regression model.

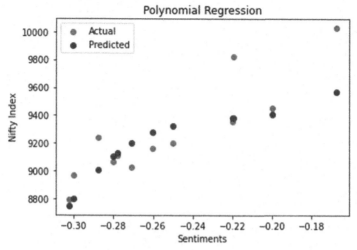

FIGURE 11.7   Plot for polynomial regression.

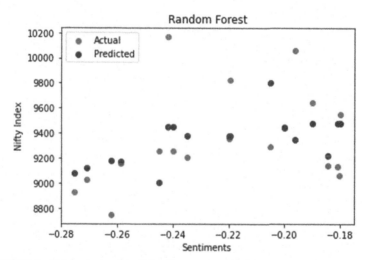

FIGURE 11.8   Plot for random forest algorithm.

11.4.4   Gradient boost regressor

The proposed method with random forest is able to produce decent accuracy; however, in order to optimize this process, this chapter uses the gradient boost algorithm, which unlike random forest, will create the trees one after the other, hence reducing the error produced. The gradient boost regression model is obtained from the sklearn package as "sklearn.ensemble. GradientBoostingRegressor" and the dataset is divided into training and test

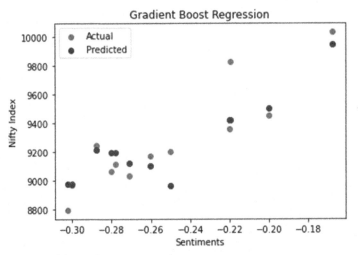

FIGURE 11.9   Plot for gradient boost regression.

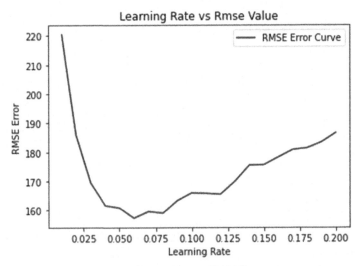

FIGURE 11.10   Change in RMSE with learning rate. *RMSE*, Root mean squared error.

data in the ratio of 80:20. The gradient boost model produces the RMSE of 157.246 and R^2 value as 0.786. Fig. 11.9 shows the plot between the sentiment value and the predicted Nifty index value for the gradient boost regression model.

RMSE is plotted against various learning rates to obtain the best fit for the model. As seen in Fig. 11.10, keeping the learning rate as "0.06" gave the least RMSE and was therefore chosen.

11.5 Conclusion

This chapter is focused on analyzing and predicting the impact of the coronavirus pandemic and its subsequent lockdown on the stock markets and indirectly on the global economy. For this case study, Nifty index values for India are considered. In order to analyze the impact overall, the chapter approaches analyzing the news headlines in the months of the pandemic. The sentiment score for each news headline is then calculated using the lexicon method. To culminate a database with the closing index values the sentiment scores for all headlines are averaged on the given date. To accommodate the COVID-19 medical scenario, this work has also created a custom dictionary and replaced words in order to fit the language of this pandemic. This database now consists of fields such as the Date, AvgSentimentScore, and the NiftyIndex. Now, regression algorithms such as linear regression, polynomial regression, random forest regression, and gradient boosting algorithm are applied to predict the Nifty index values given the headlines. In conclusion, this work proposes to use lexicon method for the initial sentiment analysis of the news headlines and then machine learning approach for predicting the Nifty index values. This work observes that the gradient boosting algorithm performs the best prediction as compared to the others for the prediction of the Nifty score.

This chapter strongly believes that this task of prediction of the stock market indexes has an immense potential to be beneficial in the future. The dataset used to train the proposed model was relatively small since it covers only a few months of this pandemic. In order to obtain a better prediction accuracy, a larger dataset can be used. This proposed method can be used to predict the stock market prices of the shares of a concerned company as well by using the appropriate historical data of the company's stock prices. Another recommendation is to extract and categorize articles that focus on financial news only, to correlate the financial sentiment and market more accurately.

References

BBC News (2013). AP Twitter account hacked in fake 'White House blasts' posts, April 24.

Bhat, M., Qadri, M., Beg, N.-u-A., Kundroo, M., Ahanger, N., & Agarwal, B. (2020). Sentiment analysis of social media response on the Covid19 outbreak. *Brain, Behavior, and Immunity*, *87*, 136–137. Available from https://doi.org/10.1016/j.bbi.2020.05.006, Elsevier.

Elliott, L. (2019). *Nations must unite to halt global economic slowdown, says new IMF head*, The Guardian, October 8,. ISSN 0261–3077.

Georgieva, K. (2020). *The great lockdown: Worst economic downturn since the great depression*, International Monetary Fund Press Release, Issue No. 20/98, March 23.

Im, T. L., San, P. W., On, C. K., Alfred, R., & Anthony, P. (2014). Impact of financial news headline and content to market sentiment. *International Journal of Machine Learning and Computing*, *4*(3), 6. Available from https://doi.org/10.7763/IJMLC.2014.V4.418.

International Labour Organisation (2020). *COVID-19 causes devastating losses in working hours and employment*, International Labour Organisation Press Release, April 7.

Joshi, K., Bharathi, H. N., Prof., & Rao, J., Prof. (2016). Stock trend prediction using news sentiment analysis. *International Journal of Computer Science & Information Technology (IJCSIT)*, 8(3), 10. Available from https://doi.org/10.5121/ijcsit.2016.8306, arXiv.

Kalra, S. & Prasad J. S. (2019). Efficacy of news sentiment for stock market prediction, In *International conference on machine learning, big data, cloud and parallel computing (COMITCon)*, 91−496. doi:10.1109/COMITCon.2019.8862265, IEEE.

Khedr, A. E., Salama, S. E., & Yaseen, N. (2017). Predicting stock market behavior using data mining technique and news sentiment analysis. *International Journal of Intelligent Systems and Applications (IJISA)*, 9(7), 22−30. Available from https://doi.org/10.5815/ijisa.2017.07.03, MECS.

Kirange, D. & Deshmukh, R. (March 2016). *Sentiment analysis of news headlines for stock price prediction*. Vol 5, Issue No 3, 5. doi: 10.13140/RG.2.1.4606.3765.

Li, X., Xie, H., Chen, L., Wang, J., & Deng, X. (2014). News impact on stock price return via sentiment analysis. *Knowledge-based Systems*, 69, 14−23. Available from https://doi.org/10.1016/j.knosys.2014.04.022, Elsevier.

Mankar, T., Hotchandani T., Madhwani M., Chidrawar A., & Lifna C. S. (2018). Stock market prediction based on social sentiments using machine learning. In *International conference on smart city and emerging technology (ICSCET)*, 1−3. doi:10.1109/icscet.2018.8537242, IEEE.

Prasad, J. (2020, June). *Covid-19 India news headlines dataset*, Version 1. Retrieved June 30, 2020 from https://www.kaggle.com/jayantiprasad/covid19-india-news-headlines-dataset.

Rohk. (2017, December). *India headlines news dataset*, Version 2. Retrieved June 27, 2020 from https://www.kaggle.com/therohk/india-headlines-news-dataset.

Shah, D., Isah, H., & Zulkernine, F. (2018). *Predicting the effects of news sentiments on the stock market.*, 4. doi: 10.1109/BigData.2018.8621, IEEE.

Souma, W., Vodenska, I., & Aoyama, H. (2019). Enhanced news sentiment analysis using deep learning methods. *Journal of Computational Social Science*, 33−46. Available from https://doi.org/10.1007/s42001-019-00035-x, Springer.

Strauss, V. (2020). *Schools of more than 90 percent of the world's students closed during this pandemic. This graphic shows how fast it happened*, Washington Post, April 6.

Further reading

Bhatnagar, V., Poonia, R. C., Nagar, P., Kumar, S., Singh, V., Raja, L., & Dass, P. (2020). Descriptive analysis of COVID-19 patients in the context of India. *Journal of Interdisciplinary Mathematics*, 1−16. Available from https://doi.org/10.1080/09720502.2020.1761635.

Harjule, P., Kumar, A., Agarwal, B., & Poonia, R. C. (2020). Mathematical modelling and analysis of COVID-19 spread in India. *Personal and Ubiquitous Computing (PAUC)*, Springer.

Kumari, R., Kumar, S., Poonia, R. C., Singh, V., Raja, L., Bhatnagar, V., & Agarwal, P. (2020). Analysis and predictions of spread, recovery, and death caused by COVID-19 in India. *Big Data Mining and Analytics*. Available from https://doi.org/10.26599/BDMA0.2020.9020013, IEEE.

Singh, V., Poonia, R. C., Kumar, S., Dass, P., Agarwal, P., Bhatnagar, V., & Raja, L. (2020). Prediction of COVID-19 coronavirus pandemic based on time series data using support vector machine. *Journal of Discrete Mathematical Sciences & Cryptography*. Available from https://doi.org/10.1080/09720529.2020.1784525.

Chapter 12

Impact of COVID-19: a particular focus on Indian education system

Pushpa Gothwal[1], Bosky Dharmendra Sharma[2], Nandita Chaube[3] and Nadeem Luqman[4]

[1]*Amity School of Enginnering and Technology, Amity University Rajasthan, Jaipur, India,*
[2]*Mayoor Private School, Abu Dhabi, United Arab Emirates,* [3]*Gujarat Forensic Sciences University, Gandhinagar, India,* [4]*Ansal University, New Delhi, India*

12.1 Introduction

The word COVID was first discovered in Ontario Cancer Institute in Toronto in 1963. Since then, various mutations were found in different parts of the world, but COVID-19, which was discovered toward the end of 2019, will be written in the history of 2020. The history indicates such incidences every 100 years. Various types of flu infections such as plague (1720), cholera (1817), Spanish flu (1918), and corona virus (2019) have been declared as pandemics. The diagonals of impact or the crater created due to the situation are major concerns today.

The novel corona virus (COVID-19) was declared a pandemic by the World Health Organization (WHO) on March 11, 2020. It is established that this virus influences the aged persons more (Zhou et al., 2020); however, this view was countered (Bhatnagar et al., 2020) and few others have done descriptive and mathematical analysis of COVID spread and made few predictions upon it which are to be observed (Harjule, Kumar, Agarwal, & Poonia, 2020; Kumari et al., 2020; Singh et al., 2020). It has globally impacted many sectors like small and large scale businesses, the world economy, health sector, transportation, wages, industries, education, etc. It is evident by the news reports and other reliable sources that this pandemic has majorly brought adverse consequences. However, it is evident that during the global lockdown, a lot of curricular activities, including regular courses, webinars, faculty development programs, lectures, training, and certification programs, have much flourished when it comes to the education sector.

Cyber-Physical Systems. DOI: https://doi.org/10.1016/B978-0-12-824557-6.00002-9

Where this online facility has made education easier and comfortable, it has its limitations also. Here, we have emphasized the impact of COVID-19 on the education sector. As per the UNESCO report, the worldwide lockdown has affected over 91% of the world's student population (UNESCO, 2019). This estimation predicts that the corona virus will adversely impact over 290 million students across 22 countries. The same report estimates that about 32 crore students are affected in India, including those in schools and colleges.

In this chapter we discuss the impact of COVID-19 with a particular focus on education. This chapter is organized as follows: Section 12.1 is introduction; Section 12.2 throws light on impact of COVID-19 on education, which has two subsections—effect of home confinement on children and teachers, and a multidimensional impact of uncertainty. Section 12.3 describes sustaining the education industry during COVID-19 and conclusions are mentioned in the last section.

12.2 Impact of COVID-19 on education

During this pandemic education sector has experienced gross changes such as a shift from regular contact classes to online platforms, modified teaching pedagogy adopted by teachers, conduction of examinations and competitive exams etc. As per the UNESCO report in the education sector, 1,190,287,189 learners have been affected and 150 countrywide closures (UNESCO, 2019). The effect of COVID-19 on the education and mental health of students and academic staff has been explored in the studies (Cao et al., 2020; Sahu, 2020). It presents some challenges due to COVID-19 on education. First, to protect the traditional teaching system, which is entirely shifted to online teaching, which requires teachers' training, strong technical support, and high-speed internet, which is not accessible for everyone. Second, the assessment and evaluation system using an online platform does not provide student performance accuracy because the originality of performance cannot be assured (Ruder, 2019). The students may use some other device to take help while answering the questions asked during the assessment. The third is the research platform, including international travel, cancellation, and postponing conferences and seminars. Other research activities have adversely affected the work (Hutton, Dudley, Horowitz-Kraus, DeWitt, & Holland, 2020). However, many such events have shifted to online platforms based on the possibilities, which has increased the participation and popularity of these events (Cao et al., 2020). The fourth concern is student mental health and career, which is grossly affected due to this outbreak (Sahu, 2020).

Studies have been conducted where the impact of COVID-19 on physicians' education was to be assessed for which they conducted seminars based on self-regulation theory and found significant results (Clark et al., 1998; Ferrel & Ryan, 2020). Ferral and Ahmad discussed the pandemic's impact due to which some hospitals in the United Kingdom canceled students' internship and observations (Ahmed, Allaf, & Elghazaly, 2020; Ferrel & Ryan, 2020). This was reassured by another study, which concluded that, as a preventive measure, many hospitals are not permitting students in hospitals, which is adversely influencing their education (Burgess & Sievertsen, 2020).

Edgar discussed the effects of COVID-19 on higher secondary education and the impact of using Science Technology Engineering and Mathematics education. In this study, the authors collected data through the telephonic mode from public school teachers, where they found a significant drop in these students' academic performance (Iyer, Aziz, & Ojcius, 2020).

COVID-19 has brought the entire education methods from traditional to online modes. There are various online platforms available for learners and professionals. The students can work with peace of mind while staying at their homes where their time, energy, and money are not wasted traveling. They are not fatigued and hence can invest themselves more in comparison to preCOVID conditions. Studying at home has also provided a more significant benefit to the students being directly monitored by parents. When it comes to theory classes, the online platform has given them a vast chance to excel. However, the practical assignments that the students are supposed to conduct in laboratories and fields have seen a major constraint. This has created a significant limitation of teaching for teachers when they cannot provide the demonstrations to the students in the absence of laboratory instruments and other necessary practical materials.

However, this has led to the timely completion of courses despite the complete lockdown but with incomplete knowledge among students whose courses are more practical. Therefore a combination of these pros and cons has brought the education world to a different level.

Several online platforms are available for lectures, training etc., which have made learning easier (Bambakidis & Tomei, 2020). However, in the absence of contact teaching, a one-to-one discussion between a teacher and students is adversely influenced. The chances of filling this lacuna are also not assured because the students will probably be deprived of contact learning before being promoted to the next level. This again leads to next level difficulties that these students may face shortly soon due to unclear concepts of previous standards/grades (Sintema, 2020). Also, in the absence of a formal class environment, the student's concentration is more likely to be adversely influenced.

Where the online facility has provided the ease of learning through flexi classes, there is no surety that the student himself or herself is attending the class. Due to network troubles, sometimes the teacher and students face many disturbances. Students sometimes get involved in mischievous activities by making fake email IDs, making noises, or giving unnecessary comments etc. The teacher faces difficulty maintaining discipline. However, this online mode is more appropriate for some disciplines than direct contact teachings, such as web designing, etc., where the practical demonstration can be better understood through online presentation and screen sharing options.

On the other hand, students from the low socio-economic class are getting no chance to experience online learning. This creates a huge and unfair social stratification where learners are left deprived of their legitimate right to education. In developing countries like India, where a huge population belongs to rural backgrounds, people are not so technology friendly. This is another challenge for the Indian education system despite the availability of technological facilities. This difficulty is faced by either or both teacher and student. Teachers who are more apt and comfortable in contact teaching cannot give their 100% through online lectures. A very advantageous and constructive aspect that emerged during the lockdown is that many professionals started throwing free online courses, training programs, workshops, webinars, etc., which have given a good chance to all the learners to update their credentials at no expense. People having busy official schedules who are usually not able to invest time in such programs are now getting a chance to upgrade themselves. On the other hand, young professionals are getting a fair chance to present themselves with more confidence.

When it comes to the physiological and cognitive effects, online education has both advantages and disadvantages. Recently, a study was conducted at Harvard Medical School on digital devices' interference in sleep and creativity. It was found that the use of digital media plays a significant role in making the neural connection for a growing human brain. However, the screen usage of more than the recommended hours can lead to lower brain development. This also leads to the disruption of sleep by undersecretion of the melatonin hormone.

Another major concern is the availability of study resources. Not all the study material is available through online mode. Several offline materials are usually available in the library but not in the online database. A student is being deprived of this material. Furthermore, the educational institutions, which have decided to conduct online examinations, face difficulty in preparing question papers. The question papers are mostly multiple choices that do not give the student a window to write descriptive answers, which are equally crucial for a student to learn. This improves the writing skills of the student.

Where the online conduct of classes and conducting examinations has its challenges, the evaluation, on the other hand, has become more convenient and transparent between the teacher and student, where the students come to know about their performance. There are platforms that allow the faculty to give online assignments and evaluation. Online teaching does not require a large infrastructure for the conduct of classes. Instead, a strong IT team is sufficient to make it workable. In direct contact teaching, the other teaching and stationary materials are required, in the absence of which teaching is likely to suffer. The online teaching platform has covered up this drawback of direct contact teaching. However, online teaching makes people more digitally dependent by reducing direct and one-to-one social interaction. This is gradually making people more technology addicts.

12.2.1   Effect of home confinement on children and teachers

Due to the COVID-19 crisis (in more than 150 countries), all levels of the education system, from preschool to tertiary education, have been affected (Bjorklund & Salvanes, 2011; Vahid, 2020), wherein gradual closure of schools and universities took place. Similar situations prevailed in the past as well, during the pandemics (Klaiman, Kraemer, & Stoto, 2011). Being confined to home or lockdown has impacted lives and livelihood across different spheres and so the education sector too, though have been able to meet the demands ensuring that via "online learning," "homeschooling," "virtual learning," or "E-learning" children's educational attainment remains undisrupted mainly (IAU, 2020).

At the tertiary level, almost all universities and colleges have offered online courses and switched to virtual lectures, classes, and webinars (Strielkowski, 2020), since digital learning has emerged as a significant aid for education from just an extracurricular facility. Although the contingencies of digital technologies rendition go past a stop-gap solution during the crisis, it has helped answer a new set of questions entirely about what, how, where, and when students shall learn. With the help of technology, students and teachers can ingress resource materials and not limit just to the text books in different formats, styles at their own pace and time by just going online. Besides teachers, smart digital technologies do not just teach only. Instead, it simultaneously observes, monitors how we study, how we learn, what interests us, the tasks that we involve in, the kind of problems that we face and find difficult to solve and adapt accordingly to meet the needs of the learner with more accuracy, specifications as compared to traditional learning within classrooms (Kumar, 2020).

However, the necessary measures taken are highly applaudable; there are various issues that arise due to prolonged school closures and home confinement (Cao et al., 2020) impacting students' well-being in COVID time wherein students feel physically less active, sleep irregularities, dietary changes marked by weight gain along with low motivation (Wickens, 2011), boredom to getting more anxious, and irritable as well. Abundant research has been carried out, suggesting having adverse effects on physical and psychological health in school-going children and students pursuing higher education at colleges and universities (Liu et al., 2019). Nevertheless, at the tertiary level, the closing of campuses left them with no choice to leave hostels and dormitories and return to their hometown; however, many got stuck too, leaving them helpless and anxious (Grubic, Badovinac, & Johri, 2020).

The switch to online education ensures minimum loss of studies suffered, and progress and attainment are also closely monitored via timely assessment and evaluations. Internal learning evaluation and assessments are considered to have high significance as it demonstrates the students' learning needs and support for taking remedial actions (Pandit, 2020). However, having been shifted to online platforms and accessed remotely, a major concern that emerged was the availability of proper internet facility networks and technology, especially in lower socio-economic zones and strata. In many countries, via online portals, TV and radio channels were started and the concern was addressed by the respective governments (Gyamerah, 2020).

Imparting of average grade points based on the course completion for students pursuing higher studies, deferring the exams till further notice, promotion to the next level using "predictive grade," were announced by few higher education institutions and schools. As per Gonzalez et al. (2020) and Black and Wiliam (2018), the evaluation method and assessment would also change from traditional high stake to small project-based and activity, assignment-based evaluation shortly as the pandemic continues. At higher education institutes, there is a hold on the ongoing research projects and field works. A virtual internship is provided and various scientific research conferences and symposiums have been postponed and canceled (Viner et al., 2020). They have moved online, whereby these virtual conferences have adversely affected networking opportunities and informal communication, creating a wide gap, especially in case of the inequalities prevailed in accessing technology to educational resources and the absence of proper remedial measures (Gjoshi & Kume, 2014).

It is perceived that higher education can be relatively managed with digital learning or remote schooling (Srivastava, 2020). As such, most of the research carried out to study the effect of the COVID-19 pandemic on education discusses the adverse effects in terms of learning and student well-being (Herold, 2017) due to home confinement and digital learning or homeschooling taking place with parental issues and concerns to provide childcare management and guidance required for their distance learning programs, availability of resources, and their socio-economic conditions (Hiremath, Kowshik, Manjunath, & Shettar, 2020).

Despite the ongoing conditions prevailing due to COVID-19, online learning has said to have long-term positive implications that can be expected in comparison to the earlier research studies that suggested that student well-being is affected by the quality of learning (Mahboob, 2020). A recent study sheds light on the significant positive impact of COVID-19 on learning efficiency and performances by adopting online learning strategies. To better understand the teaching and learning process during this crisis, it is imperative to have an education reform made to provide necessary teacher training, making further advancement of the new normal digital learning for functioning smoothly in the future as well (Stephens, Leevore, Coryell, & Pena, 2017).

Furthermore, according to WHO, COVID-19 may never be gone. Instead, people have to learn to live with it. As such, by the policymakers, distance learning is embedded in normal education, so as to help students learn coping skills to deal effectively, minimizing negative impacts in case of crises encountered.

However, as a need of the hour, education shall increasingly embrace online/virtual classrooms, keeping in mind the exposure to students' screen time in a day, planning of activities wherein parental involvement, assistance, and guidance are considered (India Today, 2020). More physical education, music, dance, home gardening along with art integration should be focussed so as to enhance creativity, and affective domains that advertently shall enhance motivation, physical activities (Sprang & Silman, 2013) and in adolescence too, continuous sitting, eye strains (Levy & Ramim, 2017) and issues like cyber bullying, video game addictions and social media browsing can be put under control. Even for university students, through distance learning, they can collaborate with others, watch lectures prerecorded, and have fruitful discussions. The lecturer can be more of a facilitator rather than an instructor. Distance learning can be as effective as a traditional face-to-face mode of learning. Students have more family time; they can engage at their own pace (Simonson, Zvacek, & Smaldino, 2019).

Moreover, there are barriers to distance learning and are unique to every country. However, its use has worldwide benefits that can be counted on, especially educating, imparting training on various focussed topics to general hobbies (Bell et al., 2017). For educators, having been faced with so many challenges to adjust and get accustomed to the distance learning platform, it is highly commendable to have done so effectively. Still, they find it convincing, and a feel-good factor also persists, as work from home has helped manage home, take care of one's self and family as mostly the time is spent on daily commuting, travels to reach the workplace, endless department meetings, colloquia or ongoing discussions on one side, and on the contrary, the research evidence (Goodman, Joshi, Nasim, & Tyler, 2015) demonstrated that parents with a low socio-economic background faced difficulties in providing nutritional meals to their children due to school closures, and also the affordability of extra-school activities compared to more advantaged backgrounds.

Nevertheless, to minimize the challenges experienced due to home confinement and school closures, distance learning should be encouraged. Need for updating with modern technology should be introduced with high-speed internet, continuous power supply, cyber security, as well as proper training to educators and students so as to have skills and competencies to operate electronic devices, along with the necessary knowledge and understanding about the method in which the information is imparted.

Clearly, due to our recent experience with the COVID-19 pandemic, many conventional academic life principles have to be reshaped. However, a common goal is being shared by all the education systems, which is to overcome the learning crisis faced and deal effectively with the COVID-19 pandemic.

While talking about the family environment, it has been observed that many faculties are reporting about online teaching difficulties. Especially in children's cases, it is reported that the families are not cooperating to maintain the class's decorum. The family members keep disturbing the child for one or the other reason, which promotes the child to continue with disturbing and inattentive behavior. The cognitive skills of the parents also have a significant role in understanding and growth of the child. If the academic and the other assignments are better understood by the parents, the children will have a constant source of support whenever needed without any delay or waiting time for the next interaction with the teacher. In this aspect, India is facing much difficulty because a large population is illiterate or less educated to compliment the contemporary educational demands of their children. Hence, the family has a central role in the learning of the child (Moon, Kim, & Moon, 2016).

12.2.2 A multidimensional impact of uncertainty

The diagonals of impact or the crater that is created due to the situation is a matter of major concern today. If we see the situation and scenario, we will find that this pandemic problem is not just medical or psychological. However, it encompasses a three-dimensional area, that is the bio-psycho-social domain of health psychology, which explains an interconnection between biology, psychology, and socio-environmental factors. This model plays an important role in defining interaction between humans and the environment and puts light on humans' interaction with their social environment in which we operate within certain domains and norms. When these domains are affected by environmental factors, a lethal combination takes birth. The world is facing the same evidence in the form of various psychological and socio-environmental outcomes, such as financial, mental health, environmental, etc. None of the areas are untouched by the pandemic influences. In the current chapter, the impacts of COVID-19 on education are explained in detail.

Suppose we see the present scenario when uncertainty is prevailing in every sector of society. In that case, it will not be superlative to say that the students of today, despite having their completed degrees, will have a certain and stable career. Such situations are making the students prone to *rumination*, which means that they are most likely to think about their uncertain future. In the present context, it is in terms of examination outcome and job security. This thought process is likely to affect their overall psyche and, in turn, will lead to a greater rise in major psychological problems.

12.3 Sustaining the education industry during COVID-19

This pandemic situation generates many education losses like postponing the board exams, competitive exams, government exams, schools and colleges closed, etc. To overcome or minimize these losses, the Human Resource Development (HRD) minister released the guideline to all educational institutes to utilize the online platforms for teaching purposes (Di Pietro, Biagi, Costa, Karpiński, & Mazza, 2020). Here, the most popular open-source of online teaching platforms are MS Team, Moodle, Zoom App, Chamilo, Webex, Canvas, Forms, Google Hangouts, and Google Meet. These platforms have helped teachers in online lecture delivery, sharing of notes, assessment, quiz conduction, etc. Several e-learning platforms are also available for students, which offer free certification or audit of the courses. These sources are Coursera, NPTEL, Swayam, edX, WHO, Harvard University, Stanford University, MITs, IITs, NITs, and many more. Therefore, in this situation, students learn at their own pace using digital platforms, while protecting themselves from the corona virus. Hence, the impact of COVID-19 on the education sector is compensated by online teaching platforms (UNESCO, 2020). This online platform also provides teachers and students with various opportunities to interact with experts as per their area of interest without any expenses. Such teaching facilitates students' effective utilization of time and more online learning activities based on their preferences.

The entire chapter can be summarized in the table mentioned below:

S. no	Pros	Cons
1	Flexible and convenient study hours.	Adjustment issues in adopting new pedagogy.
2	Novel pedagogy as a great support for sustaining education sector.	Difficulties in conducting examinations.
3	Maintaining the pace of education.	Lack of technical support and internet facility in rural and remote areas.
4	Saving time energy, resources, and money.	Originality and accuracy of performance is not assured.
5	Good for specific fields like web designing etc.	Lack of practical training resulting in decreased career opportunity.
6	People are learning technology.	No socialized learning environment.
7	Free knowledge through online courses.	Decreasing career opportunity.
8	Opportunities for new professionals.	Cognitive difficulties due to prolonged screen exposure.
9	More time to spend with family.	Physical problems like sleep difficulties, anxiety, and ophthalmological problems.
10	Multiple platforms available for study.	Psychological problems like anxiety, internet addiction etc.
11	Very convenient way of learning.	Learning rate of students affected.
12	Least resourced required for online teaching.	Due to lack of resources students are not able to get practical exposure.
13	Students can learn with own comfort.	Students not able to concentrate during class for more than 20 minutes.

12.4 Conclusion

The pandemic situation has adversely affected several sectors, but the education sector has had both advantages and disadvantages. The virus outbreak has negatively influenced other areas; the education sector has been able to sustain and has shown its advantages. Especially when we talk about digital education, it has proved to be a savior of the entire education system. However, it cannot be avoided that this digitalization has come up with its limitations. It has its pros and cons, such as home confinement, blocked socialization etc. Hence, in this epidemic situation, the fulfilment of course requirements is majorly satisfied. However, the quality of learning and outcome is adversely affected in some teaching areas, which further opens the door to more advanced education reformed by policymakers and government. Therefore, a futuristic approach to implementing such an education system needs much planning to provide a better learning platform. However, to get better results, online teaching techniques and traditional pedagogy may produce highly productive results. Therefore it can be concluded that despite having limitations, this COVID-19 pandemic has got a boost through various online platforms.

References

Ahmed, H., Allaf, M., & Elghazaly, H. (2020). COVID-19 and medical education. *The Lancet Infectious Diseases*.

Bambakidis, N. C., & Tomei, K. L. (2020). Impact of COVID-19 on neurosurgery resident training and education. *Journal of Neurosurgery*, *1*(aop), 1–2.

Bell, S., Douce, C., Caeiro, S., Teixeira, A., Martín-Aranda, R., & Otto, D. (2017). Sustainability and distance learning: A diverse European experience? *Open Learning: The Journal of Open, Distance and E-Learning*, *32*(2), 95–102.

Bhatnagar, V., Poonia, R. C., Nagar, P., Kumar, S., Singh, V., Raja, L., & Dass, P. (2020). Descriptive analysis of COVID-19 patients in the context of India. *Journal of Interdisciplinary Mathematics*, 1–16. Available from https://doi.org/10.1080/09720502.2020.1761635.

Bjorklund, A. & K. Salvanes (2011). Education and family background: Mechanisms and olicies. In E. Hanushek, S. Machin, & L. Woessmann (Eds.), *Handbook of the Economics of Education*, Vol. 3.

Black, P., & Wiliam, P. (2018). Classroom assessment and pedagogy. Assessment in education: Principles. *Policy & Practice*, *25*(6), 551–575.

Burgess, S. & Sievertsen, H.H. (2020). Schools, skills, and learning: The impact of COVID-19 on education. VoxEu.org, 1.

Cao, W., Fang, Z., Hou, G., Han, M., Xu, X., Dong, J., & Zheng, J. (2020). The psychological impact of the COVID-19 epidemic on college students in China. *Psychiatry Research*, *287*. Available from https://doi.org/10.1016/j.psychres.2020.112934, Article 112984.

Clark, N. M., Gong, M., Schork, M. A., Evans, D., Roloff, D., Hurwitz, M., & Mellins, R. B. (1998). Impact of education for physicians on patient outcomes. *Pediatrics*, *101*(5), 831–836.

Di Pietro, G., Biagi, F., Costa, P., Karpiński Z., & Mazza, J. (2020). The likely impact of COVID-19 on education: Reflections based on the existing literature and recent international datasets, JRC Technical Reports. Available at: https://publications.jrc.ec.europa.eu/repository/handle/JRC121071.

Ferrel, M.N., & Ryan, J.J. (2020). The impact of COVID-19 on medical education. Cureus.

Gjoshi, R., & Kume, K. (2014). Research on the administrator professional training and its role in the implementation of educational institutions reforms in Kosovo. *Interdisciplinary Journal of Research and Development, 1*, 26–30.

Gonzalez, T., de la Rubia, M.A., Hincz, K.P., Comas-Lopez, M., Subirats, L., Fort, S., & Sacha, G. M. (2020). Influence of COVID-19 confinement in students' performance in higher education.

Goodman, A., Joshi, H., Nasim, B., & Tyler, C. (2015). Social and emotional skills in childhood and their long-term effects on adult life. https://www.eif.org.uk/report/social-and-emotional-skills-in-childhood-and-their-longterm-effects-on-adult-life.

Grubic, N., Badovinac, S., & Johri, A. M. (2020). Student mental health in the midst of the COVID-19 pandemic: A call for further research and immediate solutions. *International Journal of Social Psychiatry, 66*(5), 517–518. Available from https://doi.org/10.1177/0020764020925108.

Gyamerah, K. (2020). *The impacts of COVID-19 on basic education: How can Ghana respond, cope, and plan for recovery?* https://schoolofeducation.blogs.bristol.ac.uk/2020/03/31/the-impacts-of-covid-19-on-basic-education-how-can-ghana-respond-cope-and-plan-for-recovery Accessed 01.04.20.

Harjule, P., Kumar, A., Agarwal, B., & Poonia, R. C. (2020). *Mathematical modeling and analysis of COVID-19 spread in India. Personal and Ubiquitous Computing (PAUC).* Springer.

Herold B. (2017). Technology in education: An overview. https://www.edweek.org/ew/issues/technology-in-education/.

Hiremath, P., Kowshik, C. S. S., Manjunath, M., & Shettar, M. (2020). COVID 19: Impact of lockdown on mental health and tips to overcome. *Asian Journal of Psychiatry, 51*, 102088. Available from https://doi.org/10.1016/j.ajp.2020.102088, 2 pages.

Hutton, J. S., Dudley, J., Horowitz-Kraus, T., DeWitt, T., & Holland, S. K. (2020). Associations between screen-based media use and brain white matter integrity in preschool-aged children. *JAMA Pediatrics, 174*(1), e193869. Available from https://doi.org/10.1001/jamapediatrics.2019.3869.

IAU (2020). *The impact of COVID-19 on higher education worldwide. Resources for Higher Education Institutions.* International Association of Universities. Retrieved from: https://www.iau-aiu.net/IMG/pdf/covid-19_and_he_resources.pdf.

India Today (2020). *Effect of Covid-19 on campus: Major steps being taken by Colleges to keep education going.* https://www.indiatoday.in/educationtoday/featurephilia/story/effect-of-covid-19-on-campus-steps-taken-by-colleges1668156-2020-04-17.

Iyer, P., Aziz, K., & Ojcius, D. M. (2020). Impact of COVID-19 on dental education in the United States. *Journal of Dental Education, 84*(6), 718–722.

Klaiman, T., Kraemer, J. D., & Stoto, M. A. (2011). Variability in school closure decisions in response to 2009 H1N1: A qualitative systems improvement analysis. *BMC Public Health, 11*, 73. Available from https://doi.org/10.1186/1471-2458-11-73.

Kumar, D.N.S. (2020). *Impact of Covid-19 on higher education.* Higher Education Digest. https://www.highereducationdigest.com/impact-of-covid-19-on-higher-education.

Kumari, R., Kumar, S., Poonia, R. C., Singh, V., Raja, L., Bhatnagar, V., & Agarwal, P. (2020). *Analysis and predictions of spread, recovery, and death caused by COVID-19 in India. Big Data Mining and Analytics.* IEEE. Available from https://doi.org/10.26599/BDMA.2020.9020013.

Levy, Y. & Ramim, M.M. (2017). The e-learning skills gap study: Initial results of skills desired for persistence and success in online engineering and computing courses. In *Proceeding of the Chais 2017 Conference on Innovative and Learning Technologies Research* (pp. 57–68).

Liu, C. H., Stevens, C., Wong, S. H., Yasui, M., & Chen, J. A. (2019). The prevalence and predictors of mental health diagnoses and suicide among US college students: Implications for addressing disparities in service use. *Depression and anxiety, 36*(1), 8–17.

Mahboob, A. (2020). Education in the time of COVID-19. Available at: http://www.flcgroup.net/courses/education101-intro/.

Moon, J. H., Kim, K. W., & Moon, N. J. (2016). Smartphone use is a risk factor for paediatric dry eye disease according to region and age: A case control study. *BMC Ophthalmology, 16* (1), 188.

Pandit, S. (2020). *Sankatma nirantar sikai.* Gorkhaparta (May 07). Available at: https://gorkha-patraonline.com/education/2020-05-06-13805.

Ruder, D.B. (2019). Retrieved from https://hms.harvard.edu/news/screen-time-brain on June 16, 2010.

Sahu, P. (2020). Closure of universities due to coronavirus disease 2019 (COVID-19): Impact on education and mental health of students and academic staff. *Cureus, 12*(4).

Simonson, M., Zvacek, S. M., & Smaldino, S. (2019). *Teaching and Learning at a Distance: Foundations of Distance Education 7th Edition.* IAP.

Singh, V., Poonia, R. C., Kumar, S., Dass, P., Agarwal, P., Bhatnagar, V., & Raja, L. (2020). Prediction of COVID-19 coronavirus pandemic based on time series data using support vector machine. *Journal of Discrete Mathematical Sciences & Cryptography.* Available from https://doi.org/10.1080/09720529.2020.1784525.

Sintema, E. J. (2020). Effect of COVID-19 on the performance of grade 12 students: Implications for STEM education. *Eurasia Journal of Mathematics, Science and Technology Education, 16*(7), 1851.

Sprang, G., & Silman, M. (2013). Posttraumatic stress disorder in parents and youth after health-related disasters. *Disaster Medicine and Public Health Preparedness, 7,* 105–110. Available from https://doi.org/10.1017/dmp.2013.22.

Srivastava, P. (2020) *COVID-19 and the global education emergency.* https://oxfamblogs.org/fp2p/covid-19-and-the-global-education-emergency Accessed 18.03.20.

Stephens, D., Leevore, M., Coryell, D., & Pena, C. (2017). *Adult education-related graduate degrees: Insights on the challenges and benefits of online programming.*

Strielkowski, W. (2020), *COVID-19 pandemic and the digital revolution in academia and higher education.* Preprints 2020, 2020040290, https://doi.org/10.20944/preprints202004.0290.v1.

UNESCO 2019, <https://en.unesco.org/covid19/educationresponse> Accessed 29.0520.

UNESCO 2020, <https://en.unesco.org/news/covid-19-crisis-Sheds-light-need-new-education-model> Accessed 29.06.20.

Vahid, F. (2020). *A message from a professor to fellow professors and students about at-home learning during COVID-19.*

Viner, R.M., Russell, S.J., Croker, H., Packer, J., Ward, J., Stansfield, C., & Booy, R. (2020). *School closure and management practices during corona virus outbreaks including COVID-19: A rapid systematic review.* The Lancet Child & Adolescent Health.

Wickens, C. M. (2011). The academic and psychosocial impact of labor unions and strikes on university campuses. In M. E. Poulsen (Ed.), *Higher education: Teaching, internationalization and student issues* (pp. 107–133). Nova Scotia Publishers.

Zhou, F., Yu, T., Du, R., Fan, G., Liu, Y., Liu, Z., ... Cao, B. (2020). Clinical course and risk factors for mortality of adult in patients with COVID-19 in Wuhan, China: A retrospective cohort study. *The Lancet, 395*(10229), 1054–1062. Available from https://doi.org/10.1016/s0140-6736(20)30566-3.

Chapter 13

Designing of Latent Dirichlet Allocation Based Prediction Model to Detect Midlife Crisis of Losing Jobs due to Prolonged Lockdown for COVID-19

Basabdatta Das[1], Barshan Das[2], Avik Chatterjee[1] and Abhijit Das[3]
[1]Department of MCA, Techno India Hooghly, Hooghly, India, [2]Department of MCA, Banaras Hindu University (BHU), Varanasi, India, [3]Department of Information Technology, RCC Institute of Information Technology (RCCIIT), Kolkata, India

13.1 Introduction

Midlife crisis (MLC), included in the dictionary first in the 60s, indicates the psychiatric symptom of lack of self-dignity and self-confidence that can happen to humans at the middle age. Although we cannot strongly say that this self-humiliation phase is a typical phenomenon, there may be many changes in life and stressors that cause midlife emotional crisis. As per the studies and researches performed in medical science, people suffering from this syndrome are prone to react differently. The fatigue, anxiety, stress, and many other psychological factors of a mid-40 urban person who is highly attached in the virtual world may lead him or her to exaggerate on the social platform. Social concern makes us aware of the fact that urban life is comparatively busier and engulfed in exertions. It is also evident that people are socially disconnected in real life and super active in electronic social media in the present time frame. The present effect of the pandemic situation and the long-term lockdown has made its impact on human society. People are pushed in an unknown atmosphere where they face death, fear of uncertainty, and even fear of unemployment. People of different ages have their psychological syndrome. With Natural Language Processing (NLP) methods and sentiment analysis, we can trace these victims who are suffering from depression. We can find several research papers in the field of sentiment

Cyber-Physical Systems. DOI: https://doi.org/10.1016/B978-0-12-824557-6.00003-0
219

analysis, which show us paths to identify depressive behavior among microblog users. We experience that there is a lack of systematic and accurate algorithms in finding proper depressive disorder.

Our contribution: Surprisingly, recognizing symptoms of MLC using sentiment analysis and NLP and finding an optimal methodology to solve this problem is yet to achive. We aim to formulate a highly efficient mechanism that will detect depressive sentences more accurately. In our work, we try to formulate an optimal mechanism implementing the concept of topic modeling with the help of the well-known theory of Latent Dirichlet Allocation (LDA), which will help us to detect those users of social platforms who are expectedly suffering from common depression and the problem of MLC.

13.2 Literature survey

Extracting snippets from a sentence collected from social media such as twitter and finding its polarity is the most obvious way of analyzing a user's sentiment. These can be compared with some predefined threshold values, which will trigger a depressive person's resultant. We mainly focus on twitter data as we can get a huge dataset from here very easily. There are several opensource tools to collect and preprocess twitter data. Gaikar, Chavan, Indore, and Shedge (2019) thus proposed a hybrid method of depression detection using SVM (Support Vector Machine) and Naïve Bayes classification. Similar work is done by Uddin, Bapery, and Mohammad Arif (2019) on purely Indian and subcontinental backgrounds. They selected several parameters and proceeded further step-by-step by tuning all the parameters. We can see that causes of the MLC can be quite different for males and females. Some recent researches, as Park and Lee (2002); Yoo, Kim, and Kim (2003); Wong, Awang, and Jani (2012), point out the causes of this psychological phenomenon among women of different geographical regions. We can find a couple of other depression detection mechanisms by Giuntini, Cazzolato, and dos Reis (2020). X Tao uses Twitter data as a strong platform for its experiment to produce a multistage detection mechanism (Tao et al., 2019). Also, in Biradar and Totad (2019); Ziwei and Chua (2019); Orabi, Buddhitha, Orabi, and Inkpen (2018); Kumar, Sharma, and Arora (2019), and Islam, Kabir, and Ahmed (2018), depressive phrases are detected from social platforms, especially from Twitter data by Biradar et al., Orabi et al., Bernice et al., Kumar et al. and Islam et al. Cacheda, Fernandez, Novoa, and Carneiro (2019) have applied the random forest technique to find depressive phrases. Another multimodal approach can be found in Shen et al. (2017) for the same problem domain.

Depression at the age of 40 can occur due to many reasons. Wethington (2000) in her work shows general medical issues of middle-aged person. Acute suffering from this problem may even lead to severe health damage: even death. In Waskel (1995), we get a glimpse of it where Sherley has described the fact of MLC and relation between depression and death. All the above literature is from medical science.

Apart from all the general cases of depressive disorder in human psychology, the recent outbreak of the COVID-19 virus has created a shiver worldwide. People are afraid of ill health, death, economic slowdown, recession, job loss, anxiety for self-health and family health, and many more. This has been geared up by long-term home quarantine and isolation from society. One such case study has been done in Shevlin et al. (2020), taking sample data from the UK people. Similar studies have been done by Aldwin and Levenson (2001) and Huang & Zhao (2020) discussing mental health at the age after 40. Mamun (Mamun & Griffiths, 2020) has shown the effect of COVID-19 among the people of Bangladesh in his very recent work. Depression caused by COVID-19 can be also seen among people of other countries. From the work of Liu, Zhang, Wong, Hyun, and "Chris" Hahm (2020) and Grover et al. (2020) we can conclude to the fact that people are getting more depressive in the span of lockdown. Visualizing and experiencing the current scenario worldwide, we have tried to apply some methods to understand the status of mental health of a certain group of humans irrespective of geographical location or financial status. As the present world is more virtual and less physical, we have opted for a social platform to analyze. The area of machine learning is the most appropriate for this study. We prefer topic modeling (Canini et al., 2009; Anandkumar et al., 2012), as it is a well-known tool for understanding and segregating sentiments. The modeling using LDA, as described by Blei (Blei, Ng and, Jordan; Blei, Ng, & Jordan, 2003), is the most relevant and easy process that uses joint probability (Newman, Smyth, Welling, & Asuncion, 2008) to specify the distribution and categorize words in different topics. Then, these topics are interlinked with separate documents where we can explain the sentiments in the percentage of topics. The entire process works on the Bayesian network (Newman et al., 2008). With this mechanism's help, machine learning and sentiment analysis become more efficient in human behavior prediction and forecasting shortly.

Reviewing all the relevant research documents, we became astonished that the sector of the mental phase of middle-aged people is still untouched concerning sentiment analysis and socioeconomic prediction. We aim to target this issue through remote analysis using topic modeling. Rest of the part of the chapter describes the problem formulations and solutions to them. The section Methodology describes in brief the actual problem area in several subsections. First we describe the symptoms of MLC and then we design the prediction model. Next we show the results and discuss on it. Following this part we draw a conclusion and also mention the future scopes of this problem.

13.3 Methodology

While we were going through news relating to COVID-19, we experienced that people all around the world are not only affected by the virus physically, but they have been mentally affected also. Mankind is suddenly put into total confinement. Surely it is very hard to change one's lifestyle in just one night or within a couple of hours. Initially, we all were thinking about the severity of the pandemic. We were managing to cope with limited resources, but gradually as the lockdown grew longer and longer, people started to exaggerate. There is certainly no need of research to understand that humanity is compelled to undergo a mental illness. Numerous documents are there to prove this statement also. However, as we all know that different age groups have different mental maturity and stability, certainly different people from separate age groups express their feelings separately. People falling under the age group $30 - 60$ are outstanding in the context that they are already at a threshold of their lives. The sudden outbreak of this pandemic has created a full nonidentical expression for them. They are mature enough to analyze the long-term consequence of this outbreak, and that single factor has triggered unparalleled feelings in them. They are scared of the disease, the economic shutdown, which is evident due to limitations created by lockdown, long-term social impacts, and many more. As a result, they engulf in the sea of depression, resulting in other physical illnesses. However, their sole feeling is hard to identify because often, people from this age group neglect their mental health and sublimely express their inner feelings. In our study, we have tried to identify this particular case of depression through the tweets and make a temporal analysis of this factor. We have taken the help of simple generative probabilistic model LDA by first understanding the theory behind it and then applying it.

13.3.1 Distinguishing midlife crisis symptoms

An MLC is not an official diagnostic. Moreover, that is the reason that we cannot tabularize the symptoms and remedies for them. Clinical researches are performed based on person-to-person questioner to a significant amount (Liu et al., 2020). It is often seen that abrupt change in behavior can be seen in the form of negligence in self-hygiene, mood swings, weight gain or loss, change in sleep habits, irritation, feeling of distress, fatigue, loneliness, suicidal tendency, and many more. Women have a biological reason for this crisis period (Yoo et al., 2003). All these emotions are likely to reflect a person's daily life. Even what he does in his daily routine or what he writes in the social platform will silently show up his condition. Adding to all these already existing facts, this sudden outbreak of COVID-19 introduced thousands of new thoughts in peoples' minds (Grover et al., 2020; Cullen, Gulati, & Kelly, 2020). A man in his mid-40s has begun to think differently to match his lifestyle with this new danger. With his knowledge and experience, he has started to think about his family's financial security, fear of job loss, health hazards of the elders of his family, and the future of the children. He has started to find ways to avoid all the risks.

Moreover, urban life has been confined in the four walls, creating mental pressure on the inhabitants. They are being obliged to stay at home doing nothing much and being super active in virtual walls. All these factors stimulate depression slowly and silently in everyone's mind, but people in midlife being more prone to the illness.

13.3.2   Designing of the prediction model

We prefer twitter data as our document's base because this platform is the most convenient medium to have a large set of data. It is obvious that whatever is posted related to COVID-19 is not of depression, nor they all post affairs related to job loss or recession. There are numerous posts of other words relating to other issues also. We try to concentrate on the fact that every tweet must not explicitly mention the relevant topics; we have to figure them out. Certainly, a single tweet will relate to multiple keywords, and one word may be categorized in different topics. We want to pull those posts made by people of age $30 - 50$ years. We first try to group the words into relevant topics, such as "depression," "pessimism," "death," "job loss," "optimistic," "joy" and obviously, "COVID-19," and "corona virus." Then the allocation method is formulated for the words that may fall under each category above. Here cross-reference happens. The sampling documents we take are the tweets in a short period in the outbreak of COVID-19 and midway of lockdown. The sampling after that is performed following Poisson distribution and Dirichlet distribution.

The prediction model is described as follows:

1. Load n tweets from neutral tweet dataset.
2. Load n tweets from depressive tweet dataset.
3. Merge neutral tweet dataset and depressive tweet dataset and shuffle the tweets.
4. Cleanse tweet dataset, that is, expand contractions, remove emojis, remove punctuations, filter stop words.
5. Make bag of words corpus from tweet dataset with most frequent words.
6. Create LDA model from bag of words corpus.

13.3.3   Application of LDA and statistical comparison

LDA is the simplest and compact way of topic modeling. The whole process follows the plate notation, referred by Blei (Blei et al., 2003; Blei, Ng and, Jordan). To train our model, we need to formulate the bag of words very carefully to get the desired output.

13.3.3.1   Formulation of Dirichlet distribution

In order to train our model, we need to choose our posterior probability and prior probability values. As the dataset is the bag of words, which form a multinomial distribution, we need a conjugate prior to this distribution. Therefore we formulate the Dirichlet distribution, which is conjugate to the multinomial distribution. To speak specifically, Dirichlet distribution is a distribution of beta distribution, and also can be derived from gamma distribution. The parameters α and β (Blei, Ng and, Jordan) are tuned so that we can filter our dataset to have more precise results. In order to do this, we follow the equation Eq. (13.1).

$$\text{Dir}(p; \alpha) = \frac{1}{B(\alpha)} \prod_{t=1}^{|\alpha|} p_t^{\alpha_t - 1} \qquad (13.1)$$

where the normalizing constant is the multinomial beta function, which can be expressed in terms of gamma function (Blei, Ng and, Jordan) in Eq. (13.2)

$$B(\alpha) = \frac{\prod_{i=1}^{|\alpha|} \Gamma(\alpha_i)}{\Gamma(\sum_{i=1}^{|\alpha|} \alpha_i)} \qquad (13.2)$$

In application, we need to find a simple generative model, which may determine that the word w_i ($i \in 1, 2, \ldots, n$) from the searched tweet is a depressive tagged word and reflects job loss. Our model must assign nonzero probability to w_i. Besides, it should also satisfy exchangeability. Each word is assigned a unique integer $x \in [0, \infty)$, and C_x is the word count in the Dirichlet process.

The probability that the word w_{i+1} is depressive is

$$\frac{C_x}{(C + \alpha)}$$

and the probability that the word w_{i+1} is an optimistic one is

$$\frac{\alpha}{C + \alpha}$$

13.3.3.2   Categorization in Bayesian model

In Bayesian probability theory, one of the joint probability events is the hypothesis, and other event is data. We wish to examine the truth of the data, given the hypothesis. In our experiment, we have a dataset D. To examine new data, we shall observe some part of the data, while we have to assume some other part of the data. Therefore we are to find a θ so that

$$h_\theta \, (\text{observed } (w_i)) \approx (\text{assumed } (w_i)). \qquad (13.3)$$

Next, we find the likelihood of training dataset assuming that the training cases are all independent of each other.

The continuous generative model of Bayesian probability leads us to determine t such that $\sim$ N $(t | \mu_{\text{extracted\_wi}}, \sigma^2_{\text{extracted\_wi}})$; where $\theta = (\mu_{\text{depressive\_wij}},$ $\mu_{\text{nondepressive\_wij}}, \sigma_{\text{depressive\_wij}}, \cdots \cdots)$

the measurement of the probability that a word w_{ij} is depressive is

$$P(\text{depressive}|t) = \frac{P(t|\text{depressive})P(\text{depressive})}{P(t|\text{depressive})P(\text{depressive})P(t|\neg\text{depressive})P(\neg\text{depressive})}$$

$$(13.4)$$

θ is derived by applying maximum likelihood.

13.3.3.3   Concept of topic modeling

Before determining the model, we apply Latent Semantic Analysis over our testing data. For each tweet, the Latent Semantic Index formulate a word-document matrix and calculate each cell value with the value computed from the calculation of (word-topic distribution $\times$ topic importance $\times$ topic-document distribution).

After determining the value for θ, we now finally approach for the construction of our final model. A pictorial reference to our model is like Fig. 13.1.

We confine our search for tweets in a short period of time and for people with age group of 30–50 years. Twitter API (Application Programming Interface) does not reveal user profile and specifically the age of a user. So we have to pull information from third-party applications. As the most discussed issue worldwide is the coronavirus now, it is not so hard to find relevant tweets in this short span. We collect sample tweets with the keywords "COVID-19", "CORONAVIRUS," and "JOB LOSS" and feed these inputs to the model.

Mathematically we want to find the probability as below:

$$P\big(\theta_{1:M}, \, z_{1:M}, \, \beta_{1:k}|D; \alpha_{1:M}, \eta_{1:k}\big)$$

Where we work on M tweets, each tweet having N words; topics are having a distribution θ of k events; z is total topics, α and η are parameter vectors; β is distribution of words; and D is our dataset.

13.4   Result and discussion

We observe the output in several stages. First, we can view the Bag_Of_Words in the form of word cloud. The fetched word cloud shows distribution of depressive and nondepressive words, as shown in Fig. 13.2.

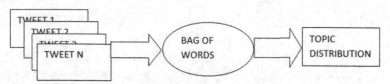

FIGURE 13.1 Simple model description of the application of latent Dirichlet allocation (LDA).

The word cloud is formed as a result of our experiment and the dataset is taken from this word cloud.

We observe the searched tweet in the next milestone, where we find the score computed by the LDA method. Each tweet, fetched singly, shows the probability of negativity after running the method in Fig. 13.3.

FIGURE 13.2 Word cloud formed in the examination using latent Dirichlet allocation (LDA).

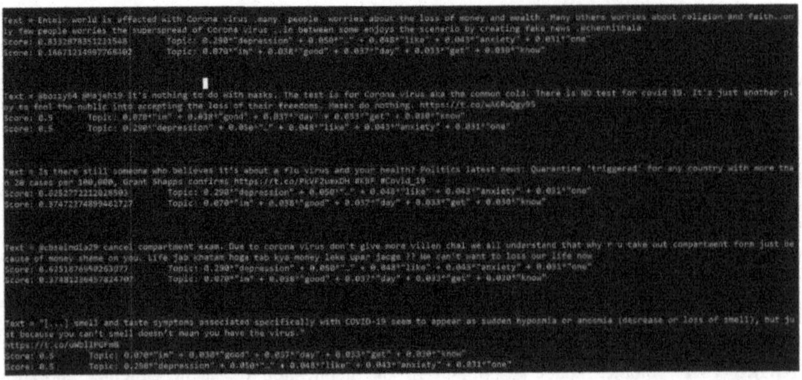

FIGURE 13.3 The probability of negativity.

Tweet#	Depression Score	Non-Depression Score
Tweet #52	0.5	0.5
Tweet #53	0.25	0.75
Tweet #54	0.58	0.42
Tweet #55	0.58	0.42
Tweet #56	0.5	0.5
Tweet #57	0.7	0.3
Tweet #58	0.83	0.17
Tweet #59	0.5	0.5
Tweet #60	0.62	0.38
Tweet #61	0.5	0.5
Tweet #62	0.83	0.17
Tweet #63	0.58	0.42
Tweet #64	0.58	0.42
Tweet #65	0.37	0.63
Tweet #66	0.82	0.18
Tweet #67	0.41	0.59
Tweet #68	0.37	0.63
Tweet #69	0.5	0.5
Tweet #70	0.62	0.38
Tweet #71	0.25	0.75
Tweet #72	0.91	0.09
Tweet #73	0.5	0.5
Tweet #74	0.78	0.22
Tweet #75	0.5	0.5
Tweet #76	0.71	0.29
Tweet #77	0.59	0.41
Tweet #78	0.72	0.28
Tweet #79	0.5	0.5
Tweet #80	0.59	0.41

FIGURE 13.4   Score chart of tweets.

For the graphical representation of the desired result, we compute the depressive and nondepressive score of tweets, as shown in Fig. 13.4.

Ultimately we get the result in the chart format shown in Fig. 13.5.

13.5 Conclusion and future scope

In recent days, many studies on the analysis and prediction of spread of COVID-19 have been performed. Some of those works were done by Poonia et al.; Kumari et al. (2020); Bhatnagar et al. (2020). After researching a minimal set of data, we have found that people of a particular age group are conscious about the pandemic situation, socioeconomic state, and income source. These facts drive them to end mental issues that they are becoming a victim of depressive disorder. This area has opened numerous scopes to

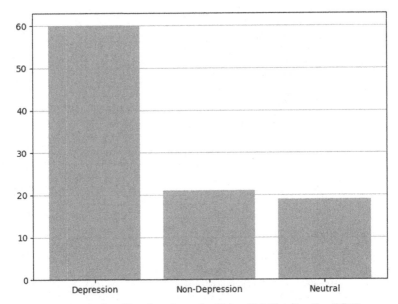

FIGURE 13.5   Graph plotted based on the result of latent Dirichlet allocation (LDA).

explore sentiment analysis (Kumar, Rama Rao, Nayak, & Chandra, 2020) using many approaches similar to the application of Probabilistic Density Functions. The present work was performed with a unigram bag of words, and future work may include an N-gram approach. There are scopes of finding a solution to the same problem domain using mixture models and EM (Expectation-Maximisation) algorithms. Also, the application of collapsed Gibbs sampling and Monte Carlo algorithm is yet to be achieved. So far, we have learned that the algorithm we used can squeeze the extensive collection of documents to shortened length, grouped according to the usage. In our next works, we shall try to optimize this property and build our algorithm for the same problem domain. Also the present approach is open to apply on a broader area of research. We shall apply this solution in finding MLC in different scenario and different time frame. At the end, we hope to build a stronger and more specific model to detect the solution of MLC in every possible domain in social media platform.

References

Aldwin, C. M., & Levenson, M. R. (2001). Stress, coping, and health at mid-life. *The Handbook of Midlife Development*, 188–214.

Anandkumar, A., Foster, D. P., Hsu, D., Kakade, S. M., Liu, Y. -K. (2012). A spectral algorithm for latent dirichlet allocation, Advances in Neural Information Processing Systems 25,

F. Pereira and C. J. C. Burges and L. Bottou and K. Q. Weinberger, 917−925, Curran Associates, Inc., http://papers.nips.cc/paper/4637-a-spectral-algorithm-for-latent-dirichlet-allocation.pdf.

Bhatnagar, V., Poonia, R. C., Nagar, P., Kumar, S., Singh, V., Raja, L., & Dass, P. (2020). Descriptive analysis of COVID-19 patients in the context of India. *Journal of Interdisciplinary Mathematics*, 1−16.

Biradar, A., & Totad, S. G. (2019). Detecting depression in social media posts using machine learning. In K. Santosh, & R. Hegadi (Eds.), *Recent Trends in Image Processing and Pattern Recognition. RTIP2R 2018. Communications in Computer and Information Science* (1037). Singapore: Springer. Available from https://doi.org/10.1007/978−981-13−9187-3_64.

Blei, D. M., Ng, A.Y., & Jordan, M. I. (2002). Latent Dirichlet Allocation, University of California, Berkeley, Berkeley, CA 94720.

Blei, D. M., Ng, A. Y., & Jordan, M. I. (2003). Latent Dirichlet Allocation. *Journal of Machine Learning Research*, *3*, 993−1022, Submitted 2/02; Published 1/03.

Cacheda, F., Fernandez, D., Novoa, F. J., & Carneiro, V. (2019). Early detection of depression: Social network analysis and random forest techniques. *Journal of Medical Internet Research*, *21*(6), e12554. Available from https://doi.org/10.2196/12554, PMID: 31199323, PMCID: 6598420.

Canini, K., Shi, L., & Griffiths, T. (2009). Online inference of topics with latent Dirichlet allocation. In Artificial Intelligence and Statistics, pp. 65−72.

Cullen, W., Gulati, G., & Kelly, B. D. (2020). Mental health in the COVID-19 pandemic. *QJM: An International Journal of Medicine*, *113*(5), 311−312. Available from https://doi.org/10.1093/qjmed/hcaa110.

Gaikar, M., Chavan, J., Indore, K., & Shedge, R. (2019). Depression detection, and prevention system by analysing tweets (March 23, 2019). Proceedings: Conference on Technologies for Future Cities (CTFC), Available at SSRN: https://ssrn.com/abstract = 3358809 or https://doi.org/10.2139/ssrn.3358809.

Giuntini, F. T., Cazzolato, M. T., dos Reis, M. D. J. D., et al. (2020). A review on recognizing depression in social networks: Challenges and opportunities. *Journal of Ambient Intelligence and Humanized Computing.* Available from https://doi.org/10.1007/s12652-020-01726-4.

Grover, S., et al. (2020). Psychological impact of COVID-19 lockdown: An online survey from India. *Dental Science—Review Article*, *62*(4), 354−362.

Huang, Y., & Zhao, N. (2020). Generalized anxiety disorder, depressive symptoms and sleep quality during COVID-19 outbreak in China: A web-based cross-sectional survey. *Psychiatry Research*, *288*, 112954. Available from https://doi.org/10.1016/j.psychres.2020.112954.

Islam, M. R., Kabir, M. A., Ahmed, A., et al. (2018). Depression detection from social network data using machine learning techniques. *Health Information Science and Systems*, *6*(8). Available from https://doi.org/10.1007/s13755-018-0046-0.

Kumar, E. R., Rama Rao, K. V. S. N., Nayak, S. R., & Chandra, R. (2020). Suicidal ideation prediction in twitter data using machine learning techniques. *Journal of Interdisciplinary Mathematics (JIM)*, *23*(1), 117−125.

Kumar, A., Sharma, A., & Arora, A. (2019). Anxious depression prediction in real-time social data (March 14, 2019). International Conference on Advances in Engineering Science Management & Technology (ICAESMT), Uttaranchal University, Dehradun, India, Available at SSRN: https://ssrn.com/abstract = 3383359 or https://doi.org/10.2139/ssrn.3383359.

Kumari, R., Kumar, S., Poonia, R. C., Singh, V., Raja, L., Bhatnagar, V., & Agarwal, P. (2020). Analysis and predictions of spread, recovery, and death caused by COVID-19 in India. Big Data Mining and Analytics, IEEE.

Liu, C. H., Zhang, E., Wong, G. T. F., Hyun, S., & "Chris" Hahm, H. (2020). Factors associated with depression, anxiety, and PTSD symptomatology during the COVID-19 pandemic: Clinical implications for United States young adult mental health. *Psychiatry Research, 290,* 113172. Available from https://doi.org/10.1016/j.psychres.2020.113172.

Mamun, M. A., & Griffiths, M. D. (2020). First COVID-19 suicide case in Bangladesh due to fear of COVID-19 and xenophobia: Possible suicide prevention strategies. *Asian Journal of Psychiatry, 51,* 102073. Available from https://doi.org/10.1016/j.ajp.2020.102073.

Newman, D., Smyth, P., Welling, M., & Asuncion, A. U. (2008). Distributed inference for latent Dirichlet allocation. *Advances in Neural Information Processing Systems,* 1081−1088.

Orabi, A. H. and Buddhitha, P., Orabi, M. H., & Inkpen, D. (2018). Deep learning for depression detection of twitter users, Proceedings of the Fifth Workshop on Computational Linguistics and Clinical Psychology: From Keyboard to Clinic, pp. 88−97.

Park, G. J., & Lee, K. H. (2002). A structural model for depression in middle-aged women. *Korean Journal of Women Health Nursing, 8*(1), 69−84. Available from https://doi.org/10.4069/kjwhn.2002.8.1.69.

Shen, G., Jia, J., Nie, L., Feng, F., Zhang, C., Hu, T., . . . Zhu, W. (2017). Depression detection via harvesting social media: A multimodal dictionary learning solution. *In IJCAI,* 3838−3844.

Shevlin, M., McBride, O., Murphy, J., Miller, J. G., Hartman, T. K., Levita, L., . . . Bennett, K. M., (2020). Anxiety, depression, traumatic stress, and COVID-19 related anxiety in the UK general population during the COVID-19 pandemic.

Tao, X., Dharmalingam, R., Zhang, J., Zhou, X., Li, L., & Gururajan, R. (2019). Twitter analysis for depression on social networks based on sentiment and stress, 2019 6th International Conference on Behavioral, Economic and Socio-cultural Computing (BESC), Beijing, China, pp. 1−4, doi: 10.1109/BESC48373.2019.8963550.

Uddin, A. H., Bapery, D., & Mohammad Arif, A. S. (2019). Depression analysis of Bangla social media data using gated recurrent neural network, 2019 1st International Conference on Advances in Science, Engineering and Robotics Technology (ICASERT), Dhaka, Bangladesh, pp. 1−6. DOI: 10.1109/ICASERT.2019.8934455

Waskel, S. A. (1995). Temperament types: Midlife death concerns, demographics, and intensity of crisis. *The Journal of Psychology, 129*(2), 221−233. Available from https://doi.org/10.1080/00223980.1995.9914960, Routledge.

Wethington, E. (2000). Expecting stress: Americans and the "Midlife Crisis. *Motivation and Emotion, 24,* 85−103. Available from https://doi.org/10.1023/A:1005611230993.

Wong, L. P., Awang, H., & Jani, R. (2012). Midlife crisis perceptions, experiences, help-seeking and needs among multi ethnic Malaysian women. *Women and Health, 52−58.* Available from https://doi.org/10.1080/03630242.2012.729557.

Yoo, E. K., Kim, M. H., & Kim, T. K. (2003). A study of the relationship among health promoting behaviors, climacteric symptoms and depression of middle-aged women. *Korean Journal of Women Health Nursing, 9*(4), 479−488. Available from https://doi.org/10.4069/kjwhn.2003.9.4.479.

Ziwei, B. Y., & Chua, H. N. (2019). An application for classifying depression in tweets ICCBD 2019: Proceedings of the 2nd International Conference on Computing and Big Data, pp. 37−41. Available from https://doi.org/10.1145/3366650.3366653.

Chapter 14

Autonomous robotic system for ultraviolet disinfection

Riki Patel[1], Harshal Sanghvi[2] and Abhijit S. Pandya[1]
[1]Florida Atlantic University, Boca Raton, FL, United States, [2]University School of Sciences, Gujarat University, Ahmedabad, India

14.1 Introduction

The worldwide fight against the COVID-19 pandemic has fathomed remarkable innovation. Governments have assumed a significant responsibility in supporting scientists to restrict the spread of crown infection and manage the current cases worldwide. Significant numbers of emergency clinics in developed nations utilize robots to help both the staff and patients.

As epidemics like Ebola, SARS, COVID-19, etc. escalate, the potential role of robots for disinfection is becoming increasingly clear. These diseases spread from person to person via respiratory droplet transfer upon close contact. The prolonged effect of the COVID-19 outbreak has introduced a severe need for disinfection while striving for continuity of work and maintaining socioeconomic functions. COVID-19 has *affected the manufacturing sector* and the economy throughout the world. This highlights the need for more automated and reliable procedures for the disinfection of large areas.

COVID-19 also spreads via contaminated surfaces. Hence, robot-controlled noncontact ultraviolet (UV) surface disinfection is being used since coronaviruses can persist on inanimate surfaces, such as metal, glass, or plastic for days. UV light devices (such as pulsed xenon ultraviolet (PX-UV)) have proved to be quite effective in reducing contamination on high-touch surfaces in hospitals. Manual disinfection requires workforce mobilization and increases exposure risk to cleaning personnel. On the other hand, autonomous or remote-controlled disinfection robots could lead to cost-effective, fast, and effective disinfection. Opportunities for innovations lie in intelligent navigation and algorithms for the detection of high-risk, high-touch areas. This, combined with other preventative measures, could help prevent the spread of any infectious disease. New generations of robots, for the macro to microscale disinfection, are under development to navigate

Cyber-Physical Systems. DOI: https://doi.org/10.1016/B978-0-12-824557-6.00011-X

high-risk areas and continually work to sterilize all high-touch surfaces (Yang et al., 2020).

Hospitals are at the top of the places to disinfect since all the sick people go for treatment. To prevent the spread of coronavirus through hospitals keeping surfaces disinfected is incredibly essential. However, this task is also dirty, dull, and dangerous considering what you can get infected with. Hence it is an ideal task for autonomous robots (Ackerman, 2020).

14.2 Background

14.2.1 Ultraviolet light for disinfection

The use of UV light has demonstrated innovation for sanitizing air, water, and instruments for more than a century. Since the 1930s, UV light has come into regular use by sanitizing medical clinics, air, water treatment in offices, microbial tainting on surfaces, etc. (Van Wynsberghe, 2016). Today UV light provides surface purification of patient rooms, restrooms, working rooms, hardware rooms, cell phones, etc. The sun is the most grounded wellspring of UV light radiation in tendency. Sun-based discharge incorporates apparent light, warmth, and UV radiation (Ackerman, 2020). Noticeable light comprises various hues that become evident in a rainbow. The UV area is additionally partitioned into four locales, namely, Vacuum (UV), Short wave (UV-C), Middle wave (UV-B), and Longwave (UV-A), as shown in Fig. 14.1. When daylight goes through the environment, UV-C and UV-B are consumed by ozone, water fumes, oxygen, and carbon dioxide (Begić, 2017). UV-A is not sifted extensively by climate. Bright light containing UV-A breaks atomic bonds inside small-scale organismal DNA, creating thymine dimers that slaughter the life forms. Bright light arriving on the earth's surface consists of roughly 95% of radiation with long frequency UV (Yang, Wu, Tai, & Sheng, 2019). It was believed that UV-A could not cause any lasting damage; however, its damaging effects on the skin were discovered over the decades. It can penetrate the skin and can cause tanning impact. Also, it often causes skin maturing and wrinkling. Several studies have demonstrated that UV-A may also enhance the development of skin cancers (Chanprakon, Sae-Oung, Treebupachatsakul, Hannanta-Anan, & Piyawattanametha, 2019; Mišeikis

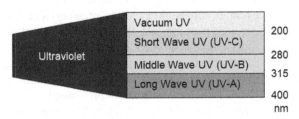

FIGURE 14.1 Ultraviolet light spectrum (nm: nanometers).

et al., 2020; Yang, Wu, Tai, & Sheng, 2019). Hence any UV disinfection system must be limited in use during human presence.

14.2.2 Exposure time for deactivation of the bacteria

To deploy a UV robot for sanitization, the exposure time for microorganisms' inactivation should be determined. An individual UV lamp bought commercially transmits a measure of UV vitality. It is essential to gauge its power by means of a force meter (Feng & Wang, 2020). The unit of brilliance is commonly communicated in a microwatt per centimeter square (μW/cm^2). The presentation time required for UV sanitization can be calculated corresponding to UV measurements related to UV portion per splendor, as shown in Eqs. (14.1) and (14.2).

$$\text{Brightness} = \frac{\text{Luminociuty (W)}}{4\pi \times \text{distance}^2(\text{cm}^2)} \tag{14.1}$$

$$\text{Time} = \frac{\text{UV dose } (\mu W\, s/\text{cm}^2)}{\text{Brightness}(W/\text{cm}^2)} \tag{14.2}$$

14.2.3 Flow chart of UV bot control logic

The flowchart in Fig. 14.2 is about disinfecting/sanitizing processes to be followed by any machine. Whenever a mechanical robot would go into a room and locate an infected area, it will start functioning according to the algorithm predefined in its memory. The steps would start from simple mobilization by the robot and disinfect the infected area in the room. If there is any kind of obstruction in the detected route then the route will be automatically changed by the machine and marched forward. But if the path is clear then it will just simply move ahead and check for further interest group in its vicinity. If it detects a movement then it will return to the docking station and shut itself down. However, if it finds no movement then it will again march for disinfection procedure and repeat the process again in a loop.

14.2.4 Calculations related to the time for disinfection

Our UV bot uses four UV lamps mounted in a roundabout fashion with a 90 degrees separation. Every light has a 15-W output. The measure of brightness within a specific separation at a given distance (35 cm in this case) can be determined using Eq. (14.3):

$$\text{Brightness} = \frac{15 \times 4(\text{W})}{4\pi \times 35^2(\text{cm}^2)} = 0.00389\,\frac{\mu W}{\text{cm}^2} = 38.9\,\frac{\mu W}{\text{cm}^2} \tag{14.3}$$

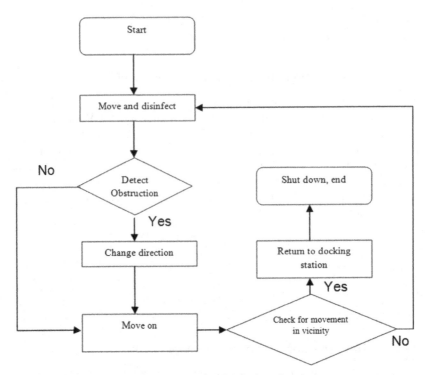

FIGURE 14.2 Flow chart of functioning of UV disinfection robot.

To compute the exposure time needed for disinfection, it is necessary to consider the UV dose and brightness. Fig. 14.5 shows the UV tower design for this project. Note that only four UV bulbs are considered since the other two faces the opposite side. This system is capable of providing a UV dose of 6600 μW s/cm^2.

Consequently, the base time required to kill germs can be calculated using Eq. (14.4):

$$\text{Time} = \frac{6,600\mu\text{W}\left(\frac{s}{\text{cm}^2}\right)}{38.9\left(\frac{\mu\text{W}}{\text{cm}^2}\right)} = 1.69s \tag{14.4}$$

14.3 Implementation

In this section, we shall discuss the architecture of the UV bot system. Fig. 14.3 shows the communication block diagram concerning Raspberry Pi connectivity to each sensor. As seen in the block diagram, the sensors provide data related to voltage and infrared reflections. The UV bot block diagram with a Raspberry Pi controller consists of a power supply, buck

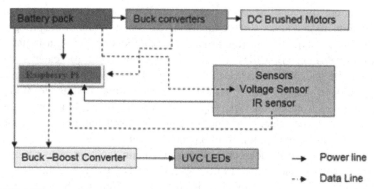

FIGURE 14.3 Interconnections between Raspberry Pi and each sensor.

converter, motors, sensors, and UV light-emitting diodes (LEDs). In our case, the buck converter is a DC-to-DC power converter, which steps down voltage (while stepping up current) from its input supply to its output load. In our design, the buck-boost converter is used as a switched-mode power supply that combines the principles of the buck converter and the boost converter in a single circuit. Our UV bot assembly consists of 10 LED modules, two direct current (DC) motors, and one Raspberry Pi, interconnected via data paths and power lines.

In the proposed model, a rechargeable three-cell lithium particle pack that self-controls the gadget has been suggested. Since this is a three-cell pack, it has a voltage supply between 9 V and12.6 V. The three-cell lithium gives us an expected limit of 13 Ah for the direct outcome possible. Since everything is similar, the motors will not be running continually at maximum speed, so a more sensible range would be 10−12 Ah.

These motors were evaluated for 12 V and 0.3 A. We are utilizing brushed DC electric motors, which are internally commutated electric motors designed to be run from a direct current power source. They were chosen because of their simplicity and ease in control concerning our robot's movement. These will be constrained by the appended buck converter, which will consider precise speed control. The basic exchanging buck-boost converter permits a DC voltage to be productively ventured down to whatever voltage is required. This is the middle station between the regulator and the DC engine. These converters were evaluated to yield at any rate 0 V−10 V +/− 10% along with the option to operate at 1 A +/− 20% gracefully.

The UV tower has been incorporated for the essential function of sterilizing the surface it encounters. UV-C is high-vitality bright radiation strong enough to destroy DNA and RNA of infecting bacteria and viruses. This permits the UV LEDs in our system to kill infectious and microscopic organisms on the surface. The greatest concern we have is to ensure that we are both splashing the surface with enough radiation to purify and keeping a huge amount of radiation from spilling out. We would utilize some control calculations to guarantee surface inclusion. To forestall radiation spillage, we would utilize a type of covering around the robot's edge as a splashguard.

In our design, six LEDs are set in an arrangement for each stack to improve the gadget's effectiveness. The issue is that the battery voltage can fluctuate over a large range. Buck-boost converters make it easy to modulate voltage in such cases. In fact, for a wide variety of popular applications, including consumer electronics, power amplifiers, self-regulating power supplies, and control applications rely on such converters.

In our design, the buck-boost converter's main function is to receive an input DC voltage and output a different level of DC voltage, either lowering or boosting the voltage as required by the LEDs. Thus our buck-boost converter acts as a control unit and functions similarly to a buck converter and boost converter, except in a single circuit. As a control unit, it senses the input voltage level and takes appropriate action on the circuit based on that voltage.

Thus the buck-boost converter assists the battery-powered system. When the system has no charge and only a certain amount of voltage to charge up the system, the "boost" portion kicks in. The buck-boost converter is used to make the input voltage produce an output voltage more significant than the input voltage. This allows the max voltage to charge the system as quickly as possible. As the system's charge reaches its max capacity, however, there is a risk of overheating. This is where the "buck" portion of the buck-boost converter kicks in and recognizes that the system's charge is getting close to full. At that point, it lowers the voltage gradually. Once the system reaches the max charge, the voltage level drops to zero.

These converters have been appraised to yield at 11 V $+/-$ 10% along with a flexibility of a maximum of 3 A $+/-$ 20%. Infrared sensors are utilized to detect the edges of tables or other similar surfaces and pass judgment if the robot is yet on the table or a raised platform. We do not want the robot to drive itself off the raised platform; accordingly, the infrared (IR) sensors have been incorporated to recognize when the robot will drive off and fall. Likewise, we can utilize these sensors to decide whether the robot has been unexpectedly lifted off the table by a human being. This is significant for well-being of persons because of the harmful effects of UV light. By having these data, the robot can stop the LEDs when it establishes that it was lifted off the table.

14.4 Model topology

In the coming decade, around 35,000,000 individual help robots will be deployed all around, as indicated by the International Federation of Robotics. These robots would not be necessarily functioning independently at homes or work environments of individuals; however, they will assist us in many situations. For example, these robots may welcome us at stores, help us with what we need to purchase, clean our floors, sterilize rooms, read stories for the children when we are away, etc. (Ahn, Liang, & Cai, 2019; Casini et al., 2019; Kumar & Sharma, 2020; Nguyen, 2020; Schmidt et al., 2019). Administration robots have a gigantic deployment scope due to their abilities such as infrared detection, degrees of self-rule, appearances (machine-like animals like humanoid), etc. As discussed earlier, the deployment of robots for disinfection has become necessary due to the COVID-19 pandemic.

When it comes to manual disinfection procedures, the quality of environmental cleaning in hospitals has revealed that mere cleaning efforts are often insufficient. Janitorial cleaning often leaves microbial contamination on surfaces due to human touch during cleaning. Hence, interest in "touchless" technologies has exploded in high-resource healthcare settings. UV light-emitting robots are often deployed to reduce environmental bioburden and decrease residual pathogenic organisms' risks in patient-care areas. Investment in robots allows healthcare centers to follow traditional manual cleaners with an automated cleaning process to ensure rooms are optimally cleaned and disinfected (Schmidt et al., 2019).

14.4.1   UV-C light robotic vehicle

Fig. 14.4 depicts the schematics of the U-bot, a robotic vehicle with the ability to disinfect using UV-C. It allows the user to deploy it in either the default mode or the programmed mode. The programmed mode allows the user to choose the cleansing activity plan based on specific prerequisites. The user can set up the splashing time, portion, and area that has to be covered, etc. using a terminal. Using the built-in mechanics, the U-bot then carries out the sanitization plan indicated by the boundary settings.

U-bot design presented in this chapter is based on a disinfecting device developed by Garcia and Garcia (Bilenko et al., 2019). Their device uses a UV light source to eradicate the medium of infestation agents such as molds, viruses, bacteria, and dust mites. This device enhances disinfection by providing mechanisms for enhanced penetration of the UV light into the cleaning medium. In the case of the U-bot design presented in this chapter, certain modifications have been carried out. The modified design consists of a light assembly with a height of 50 cm, marked as a silver colored UV-C light bulb in Fig. 14.4. This assembly is made up of six UV light bulbs, as shown in Fig. 14.5, which are inserted in the silver colored skeleton frame. The docking assembly in Fig. 14.4 also serves as the charging station for the battery (Chanprakon, Sae-Oung, Treebupachatsakul, Hannanta-Anan, & Piyawattanametha, 2019).

The silver skeleton frame is mounted on a telescopic extension bar, which can rise to 150 cm from the surface level. The 150-cm range allows the U-bot to cover the top surface of tables and other elevated platform surfaces. A safety switch may be positioned at the top, bottom, or other accessible

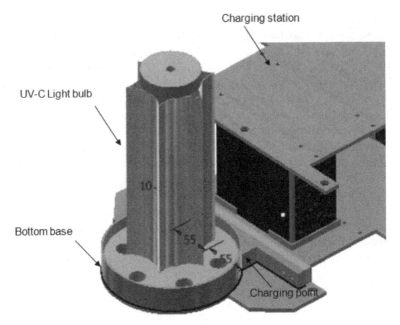

FIGURE 14.4 UV-C light robotic vehicle.

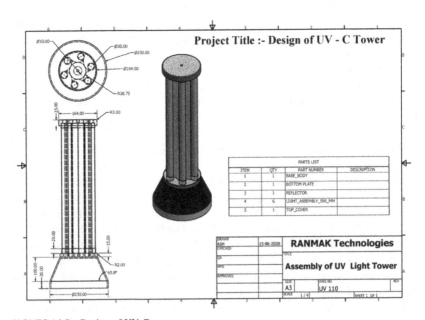

FIGURE 14.5 Design of UV-C tower.

portion of the handle assembly. The switch can be configured so that the safety switch must be depressed or activated during the device's operation to permit the light bulb to be active.

A camera (not shown in the figure) is mounted at the top of the skeleton frame, which can provide 360-degree view. It provides the data for scanning the room and operates via commands sent from the programming module. Several sensors help U-bot to identify high concentrations of infected areas that require more attention. Moreover, a couple of other wheels are incorporated into the bottom unit in order to allow it to maneuver room terrain safely. These wheels allow it to lift and navigate out of areas where it is likely to get stuck. The UV-C light, when activated, sweeps along walls, baseboards, and furniture legs.

At the bottom of the U-bot, a small IR-sensor is placed and connected to the programming module. IR sensor also functions as a motion sensor, proximity sensor, temperature sensor, electrical parameters sensor, and pathogen identification sensor. The U-bot has a six-axis gyro sensor, which is controlled by an embedded microcontroller chip. A wireless radio frequency (RF) communication assembly has been incorporated, allowing bi-directional communication with a wide range of remote accessories, sensors, and controls. Instead of manually starting the robot, you can set a schedule with the exact time each day. In that case, the U-bot will clean the specific area at the desired time automatically. You can set different times for each day, depending on your schedule.

CAD prototype of the UV-C light tower is illustrated in Fig. 14.5.

14.5 Conclusion

The design of the UV disinfection robot, working flow chart, and the structure is discussed in detail in this chapter with recent modifications and applications in the real world. COVID-19 pandemic has resulted in a dire need to address sanitization and disinfection procedures in all public places. Recent advances in robotics have resulted in the widespread use of automated systems with reduced human involvement. This chapter discussed a UV robot's design that can disinfect areas while moving through the room intelligently. These types of UV robots are getting researchers' attention while they are comparatively new in disinfection. The main challenge is related to ensuring safety besides intelligent movement algorithms and deployment of mechatronic components. The major advantage of using robotic systems is related to compliance. With human errors, there is no guarantee that the same error will not occur again. Achieving 100% compliance cost-effectively is quite tricky. On the other hand, any robotic system error will never reoccur once it has been fixed either mechanically or through the program that governs its function. Hence it is possible to expect 100% reliability after thorough testing.

References

Ackerman, E. (2020). Autonomous robots are helping kill coronavirus in hospitals. *IEEE Spectrum*.

Ahn, C., Liang, X., & Cai, S. (2019). Bioinspired design of light-powered crawling, squeezing, and jumping untethered soft robot. *Advanced Materials Technologies*, *4*(7), 1900185.

Begić, A. (2017). *Application of service robots for disinfection in medical institutions. The International Symposium on Innovative and Interdisciplinary Applications of Advanced Technologies* (pp. 1056–1065). Cham: Springer.

Bilenko, Y., Dobrinsky, A., Smetona, S., Shur, M., Gaska, R., & Bettles, T.J. (2019). United States Patent No. 10,442,704. Washington, DC: United States Patent and Trademark Office.

Casini, B., Tuvo, B., Cristina, M. L., Spagnolo, A. M., Totaro, M., Baggiani, A., & Privitera, G. P. (2019). Evaluation of an ultraviolet C (UVC) light-emitting device for disinfection of high touch surfaces in hospital critical areas. *International Journal of Environmental Research and Public Health*, *16*(19), 3572.

Chanprakon, P., Sae-Oung, T., Treebupachatsakul, T., Hannanta-Anan, P., & Piyawattanametha, W. (2019). An ultra-violet sterilization robot for disinfection. In *2019 5th International Conference on Engineering, Applied Sciences and Technology (ICEAST)* (. 1–4). IEEE.

Feng, Q. C., & Wang, X. (2020). Design of disinfection robot for livestock breeding. *Procedia Computer Science*, *166*, 310–314.

Kumar, A., & Sharma, G. K. (2020). Artificial intelligence technologies combating against COVID-19. *Dev Sanskriti Interdisciplinary International Journal*, *16*, 56–60.

Mišeikis, J., Caroni, P., Duchamp, P., Gasser, A., Marko, R., Mišeikienė, N., ... Früh, H. (2020). Lio—A personal robot assistant for human-robot interaction and care applications. *IEEE Robotics and Automation Letters*, *5*(4), 5339–5346.

Nguyen, T.T. (2020). Artificial intelligence in the battle against coronavirus (COVID-19): A survey and future research directions. *Preprint, DOI, 10*.

Schmidt, M. G., Attaway, H. H., Fairey, S. E., Howard, J., Mohr, D., & Craig, S. (2019). Self-disinfecting copper beds sustain terminal cleaning and disinfection effects throughout patient care. *Applied and Environmental Microbiology*, *86*(1).

Van Wynsberghe, A. (2016). Service robots, care ethics, and design. *Ethics and Information Technology*, *18*(4), 311–321.

Yang, G.Z., Nelson, B.J., Murphy, R.R., Choset, H., Christensen, H., Collins, S.H.,... & Kragic, D. (2020). Combating COVID-19—The role of robotics in managing public health and infectious diseases.

Yang, J. H., Wu, U. I., Tai, H. M., & Sheng, W. H. (2019). Effectiveness of an ultraviolet-C disinfection system for the reduction of healthcare-associated pathogens. *Journal of Microbiology, Immunology and Infection*, *52*(3), 487–493.

Chapter 15

Emerging health start-ups for economic feasibility: opportunities during COVID-19

Shweta Nanda

IILM Graduate School of Management, Noida, India

15.1 Introduction

The Internet of Things (IoT)/artificial intelligence (AI)-based health start-ups are building a complete health ecosystem for people worldwide, connecting patients, caregivers, doctors, and hospitals with a common thread. The smartphone was initially meant to make and receive phone calls and has now become a medium to share a rich stream of personal data on a cloud-based platform (Ferguson et al., 2016).

"Coronaviruses are a class of viruses that may cause illness in animals or humans. In humans, several coronaviruses are known to cause respiratory infections ranging from the common cold to more severe diseases such as Middle East Respiratory Syndrome and Severe Acute Respiratory Syndrome. The most recently discovered virus causes coronavirus disease known as COVID-19." Many Indians returned to work as the economy opened up gingerly; while things are turning uglier for the economy and start-ups, there are some bright sparks. Due to technology, people got abreast of this virus. Further, the following applications of AI proved its capabilities in assessing nature of risk for COVID-19

- **Monitoring hospital visitors and patients using AI**: Facial scanner and thermal scanners help identify the facial attributes and fever conditions of the people.
- **Remote monitoring**: This is a sensor-based technology. If the sensor is placed under the patient's mattress, we can track the heart rate, respiratory rate, and body movement. It helps to take care of COVID patients.
- **COVID voice detector**: The AI tool will detect the infected one by evaluating their cough and breathing problems.

Cyber-Physical Systems. DOI: https://doi.org/10.1016/B978-0-12-824557-6.00010-8

Use of IoT: IoT is playing a vital role in limiting the spread of virus. It also helps in treating COVID-19 infected people with touch-free attendance, sanitization conformity, and supervising body temprature. Now the warehouse, hospitals, and offices are using it.

- Tracking the coronavirus pandemic.
- Connected thermometers.
- Smart wearable.
- IoT buttons

There is a tremendous potential of wearable ECG systems to bring revolution in treatments for various cardiovascular diseases (CVD). Patients with chronic ailments will be benefited most by these types of wearable remote systems. IoT-based remote wearable sensors and the integrated cloud platform enable healthcare providers to capture vitals remotely and perform analysis (Nanda, Khattar, & Nanda, 2019). Continuous ECG graph monitoring of geriatric patients and offering them with preventive care at a preliminary stage would drastically reduce the aberration stances. Four major applications of wearable devices have been observed: portable devices, home adaptations, electronic systems, and connected devices (Doughty & Appleby, 2016). It is also proposed that through smartphones, consumers can obtain the most routine lab test which transfers the ownership of data from health stakeholders to the patients (Kish & Topol, 2015). The IoT-based remote health monitoring system may also be used for falling detection, elderly care, sports training, rehabilitation training, postoperative care, and other fields. Microcontrollers are used to record physiology-based data with the help of these devices. The central controller helps process the data and generates a message of warning to the person taking care and even help predict upcoming disease.

In order to maximize the benefits from IoT, many consortiums of companies and industry bodies have drafted technology and regulatory protocols to promote standardization and uniformity. IoT is benefitting both consumers and industries in distinct and innovative ways through varied applications. IoT helps create value for stakeholders through the availability of information, with the help of technologies such as sensors, networks, standards, augmented intelligence, and augmented reality (AR). It is also helping to enhance process efficiencies significantly across industries, particularly manufacturing and healthcare, thus taking industrial applications to the next level.

According to a report by Federation of Indian Chambers of Commerce and Industry (FICCI) presented by Deloitte on "Indian Medical Electronics Industry Outlook 2020," the Indian demographic factors offer the opportunity for huge growth in medical electronics due to potential demand for healthcare. The Health Ministry has launched for the elderly "National Program for the Health Care of the Elderly" (NPHCE) in India with a provision of INR 288 crores to invest for the same in 2010–2025. A major health problem that is change in disease profile—noncommunicable diseases (NCDs) is the leading cause of death. Sedentary lifestyles have led to an increase in lifestyle/NCDs such as CVD, diabetes, cancer, etc. Lifestyle diseases such as obesity, CVD, and diabetes are also forecasted to become more pervasive. The Government of India has initiated the National Program for Prevention and Control of Cancer, Diabetes, Cardiovascular Diseases, and Stroke. A growing number of medical devices are becoming potential wearables in India, including ECG monitors, glucose monitors, blood pressure monitors, and pulse oximeters.

Whereas in the developed nations, technologies such as Preventive's BodyGuardian remote monitoring system or Avery Dennison's Metria Wearable Technology are setting the stage to deliver patients data to doctors seamlessly. Bluetooth is key in systems such as 9soulutions IPCS, which uses it to track elderly patient's movement and send health measurements to caregivers. BodyTel uses Bluetooth to allow patients to send body measurements to their doctors wirelessly. Similarly, in a country like India where diabetic patients are high, continuous glucose monitors have a wide scope to monitor the glucose level in human bodies and help sustain at the desired level by injecting insulin from time to time. C8 Medisensor is a wearable product that conducts noninvasive optical glucose monitoring by transmitting a pulse of light through the skin and continuously updates the data to a smartphone via Bluetooth.

15.2   Health-tech verticals for start-ups

To control the spread of coronavirus, AI and IoT start-ups leverage the tools and solutions to help the crisis. There is a start-up called Indian Robotics, providing its robots with screening and diagnostics. These robots help to collect the data from the patients, symptoms exhibited, and validation. These robots enable a video conference with a doctor from rural locations to help out the people who are facing symptoms like high fever etc. AI and IoT are becoming primary weapons for tracking and tracing the cases.

Medikabazaar is the start-up that is providing the online B2B platform only for doctors and hospitals. The supply chain of this start-up is in Tier II cities, Tier III cities, and remote locations. They provide all medical equipment, masks, thermometers, test kits, body covers, etc. for doctors online. MyLab, Bione, Redcliffe Life Sciences are the start-ups with low-cost manufacturing ventilators for COVID patients to develop AR-based solutions in India. With respect to COVID, it has been found by Bhatnagar et al. (2020) that age is not a significant factor for a person to be infected by COVID.

Noida-based biotech start-up DNA experts have also developed testing kits, increasing the total number of tests by reducing the time taken per test. While most testing kits used in India are taking around 2.5 hours for the result, DNA experts claim its COVID test takes just 58 minutes to test a sample. The start-up is incubated at the state-run Centre for Cellular and Molecular Platforms.

In this scenario, a large number of Medtech start-ups are working on diagnostic solutions and preventive healthcare, and some as healthcare aggregators. There are two broad categories of Medtech start-ups. First, that is, harnessing technologies like AI and the IoT to change India's healthcare landscape. According to the Indian start-up ecosystem, the second category, traversing the maturity cycle report 2017, is a multiple health-tech start-up in the subverticles of medical solutions like a marketplace for health services, health lab aggregators, wellness platforms, online pharmacies, e-diagnostic and ambulance aggregators. Among the health-tech start-ups 87% are B2C and 26% are B2B.

15.3 Research gap

Their lies a huge gap between the available healthcare start-up services and the adoption by prospective customers seeking affordable healthcare solutions. This conceptual study provides a framework to understand the present real-time analytics-based healthcare start-up solutions and suitable approaches to extend their availability during COVID-19.

15.4 Aim of the study

To categorically identify the framework of healthcare start-ups in India and how real-time analytics-based start-ups can meet remotely located patients' demand.

To explore which healthcare start-ups would meet the diagnostics demand of Tiers II and III places during COVID-19.

15.5 Research methodology

15.5.1 Problem statement

This research would be exploratory, diagnostic, and conclusive. The study would be exploratory as it would explore the various real-time analytics-based health start-ups in India. The methodology used in this study is based on system thinking. The methodological framework used in this study consists of two phases. First, an attempt was made to structure the problem. Second, a causal loop model was developed to capture feedback loops to explain the system (Elias, 2019). This study was limited to qualitative modeling based on system dynamics (Cavana & Mares, 2004).

The study would also be conclusive as it would provide meaningful strategies and approaches toward a sustainable healthcare model in India. The study provides a holistic view of start-ups that have reduced healthcare service costs by leveraging technologies.

15.5.2 Type of research

In this study, the maximum start-up funding has been observed in the health-tech verticals of "tech-enabled diagnostic services" and "anomaly detection and disease monitoring." However, their availability is limited to Tier I. This research attempts to study the contributions made by real-time analytics systems to make them affordable/accessible for Tiers II and III, especially during COVID-19.

15.5.3 Secondary data

This study is conducted with secondary data available from various sources like IoT World Congress Report, Deloitte, Nasscom publications, FICCI Report, and research articles that are very relevant to this field.

15.5.4 Data analysis methods

Data analysis would be done using inferential analysis for drawing inferences and interpretations in this study. Python has been used to do extrapolation through plots and determine the potential medical devices required during COVID-19 (Tables 15.1 and 15.2).

TABLE 15.1 Healthcare start-ups based on real-time analytics/advanced analytics.

Category: Real-time analytics	Start-up name	Services
Cloud analytics	Bagmo or "Blood Bag Monitoring Device"	Blood Bag Monitoring Device monitors the temperature of the blood bag during the transportation and storage.
Genetic analytics	Prantae Solutions	EyeRA for early detection of preeclampsia, which is a pregnancy disorder. Received the CII-IPR award.
Advanced analytics	Waferchip Techno Solution	Continuous monitoring of ECG data and capturing through Bluetooth through wearable device called Biocalculus.
Cloud analytics	EzeRX	Device called AJO measures anemia, jaundice, and oxygen saturation from noninvasive IoT device.
Cloud analytics/ predictive analytics	CardioTrack	The device EMR App captures ECG data through Bluetooth, calculates average heart rate, and helps in doing predictive analysis for upcoming heart disease.

TABLE 15.2 Healthcare start-ups based on healthcare aggregator services.

Category: Healthcare aggregators	Start-up name	Services
Hyperlocal health services	Medikoe	It is a platform connecting hyperlocal market to facilitate them with healthcare and wellness services (partners include Manipal Hospitals, Jiyo Healthcare, NM Medical Diagnostics, and Positive Homeopathy).
Home healthcare	Medwell Ventures	Offering home healthcare and palliative care to middle-aged and geriatric patients.
Pharmacy aggregators	Care24	Its services include from prescription to complete recovery, nursing at home, physiotherapy at home, infant care at home etc.
B2B digital health-tech platform	Medikabazaar	It offers medical devices and machine delivery through catalog enlisting of over 25,000 products to above 20,000 pin codes including Tiers II and III hospitals.

15.6 Health-tech category I Indian start-ups

15.6.1 Heath-tech category II Indian start-ups

15.6.1.1 Inferences

Out of the two healthcare start-up categories, real-time analytics start-ups leverage cloud platforms to offer remote healthcare solutions. This category emphasizes using the products-as-a-service model and thus can make its service remotely accessible during COVID-19. They use a service-driven business model through AI/IoT healthcare solutions.

15.6.2 Variables gathered from stakeholder interviews

Further, as an analysis of our interviews with the health service providers reinforced the fact that with the introduction of IoT/AI into businesses and society, there is a promise of productivity and efficiency by improving real-time decision making, solving critical problems, and creating new innovative services and experiences in the COVID times. However, insights have been received on how Tiers II and III can leverage existing health start-up services through affordable, innovative services. They emphasized the product-as-a-service subscription model wherein the remote medical device user is charged only if they opt to share his medical data for predictive/preventive analysis. The future

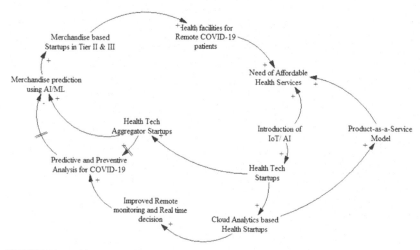

FIGURE 15.1 Causal loop model.

scope of devices are based on a subscription model for yielding recurring revenue at an affordable cost for Tiers II and III customers. However, there are very few start-ups in the healthcare aggregators category, which leverage AI/machine learning (ML) implementation in merchandizing prediction and procurement cost reduction to reach in Tiers II and III. (Fig. 15.1).

15.6.3 Causal loop model

Adaptive AI and ML technologies not only have the potential to optimize device performance in real-time but also to forecast the demand for the diagnostics devices in Tiers II and III primary healthcare centers.

Diagnostics, IV Diagnostics, and other devices have a maximum import dependency in India. Diagnostic imaging (e.g., CT scan, X-Ray, MRI, USG, X ray-tubes), IV Diagnostic (lab equipment and reagents, etc.), and other medical devices (ECG, optional equipment, heart − lung machine, etc.) form 70% of total import in India in FY1613.

To identify the future potential of the segment-wise medical devices in India, joint plots are made using python coding. It has been done to extrapolate the future demand of these medical devices and identify the most promising potential demand segments.

Coding is specified in the respective figures (Figs. 15.2, 15.3, 15.4, 15.5, 15.6, 15.7, 15.8).

Thus these plots provide the scope of opportunities in the manufacturing of the devices. As per the plots, the maximum demand potential is Diagnostic Imaging, IV Diagnostics, and others. Currently, out of the 750 medical device manufacturers present in India, a majority are Small and

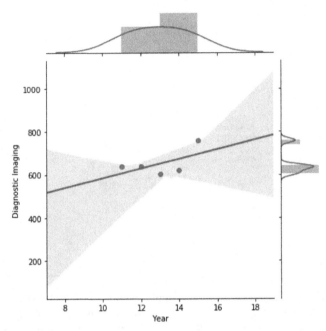

FIGURE 15.2 Diagnostic imaging plot.

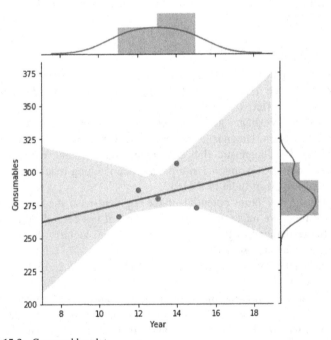

FIGURE 15.3 Consumables plot.

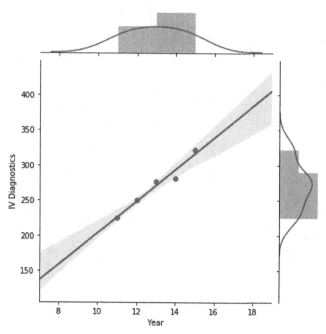

FIGURE 15.4   IV Diagnostics plot.

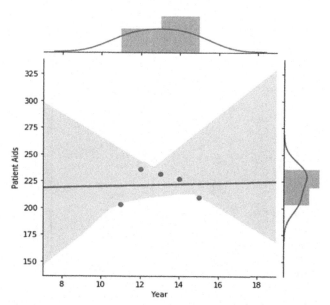

FIGURE 15.5   Patient aids plot.

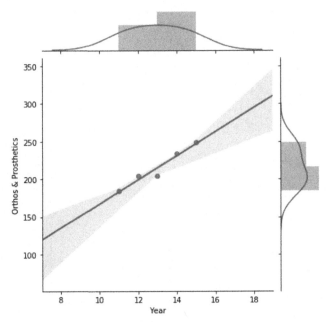

FIGURE 15.6 Orthos and prosthetics.

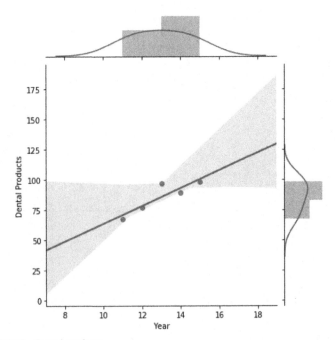

FIGURE 15.7 Dental products.

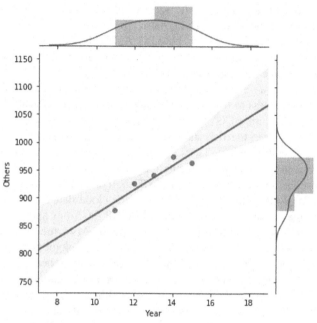

FIGURE 15.8   Other plot.

Medium Enterprises (SMEs) and Micro, Small and Medium Enterprises (MSMEs) (90% have an annual turnover of less than USD 10 million) and contribute 30% (USD 1.1 billion) to the Indian medical devices market (Deloitte, NatHealth, 2016). Simultaneously, indigenous medical devices start-up manufacturers are yet to scale up for bulk production for Tiers II and III.

15.7   Conclusions

IoT/AI in healthcare has a vast potential, but the market penetration is quite low in the Indian market. IoT/AI-enabled health start-ups have the ability to customize their services following the local demand, especially during COVID-19. The demand for mobile health solutions is driven by a remotely located population where the immediate reachability of the medical devices and services is a challenge. Despite identified COVID, vulnerable zone health-based merchandise could not reach on time. The AI/ML-based merchandise prediction solutions for diagnostics and critical care devices in Tiers II and III can reduce patients' burden to commute to big cities for immediate care.

These patients are also looking for affordable, innovative health solutions. The above study deals with the systems thinking approach to deal with the availability gap of affordable health services to the remotely located patient. In the causal loop model (Fig. 15.1), there are two reinforcing loops and one balancing loop. It tries to identify how IoT/AI/ML-based start-ups can benefit COVID-19 patients through start-ups based on cloud analytics or AI/ML-based health device merchandise providers.

The product-as-a-service solution attracts the start-ups in India with an idea to provide open access to their data collection practices and its analysis for preventive strategies during COVID-19. This model not only eases postaberration care but also reduces the risk stances by real-time monitoring/diagnosing the health patterns and generating alerts to stakeholders based on the threshold limits. However, in the model's balancing loop, a delay has been observed in the merchandise facilitating health start-ups in Tiers II and III locations.

There is a considerable scope to capture the cloud data from Tiers II and III through the seamless integration of IoT/AI/ML start-ups and clinical practitioners to perform customized medical research. Also, the critical disease profile varies with every country, and so does the supplementary supporting services of healthcare aggregators who make the core healthcare service accessible. Through the above healthcare start-up and funding initiatives, a shift in need has been observed from the healthcare category I present in Tier I to its availability/accessibility in Tiers II and III. Thus this study suggests the companies implement its proprietary AI and ML tools to report accurate stock projections for medical establishments in these areas and leverage products as a service model to make the devices affordable.

In the critical COVID-19 patients, a need was predicted for the ventilators in the remote areas. AI/ML in machine delivery and medicines supply chain/operative care has the potential to timely deliver the required critical devices. Thus there is a vast potential for start-ups manufacturing Diagnostics Imaging & IV Diagnostics devices for Tiers II and III places.

References

Bhatnagar, V., Poonia, R. C., Nagar, P., Kumar, S., Singh, V., Raja, L., & Dass, P. (2020). Descriptive analysis of COVID-19 patients in the context of India. *Journal of Interdisciplinary Mathematics*, 1–16.

Cavana, R. Y., & Mares, E. D. (2004). Integrating critical thinking and systems thinking: From premises to causal loops. *System Dynamics Review*, *20*(3), 223–235. Available from https://doi.org/10.1002/sdr.294.

Deloitte, NatHealth (2016). Medical devices making in India—A leap for Indian healthcare, available at: https://www2.deloitte.com/content/dam/Deloitte/in/Documents/life-sciences-health-care/in-lshc-medical-devices-making-in-india-noexp.pdf (accessed July 23, 2020).

Doughty, K., & Appleby, A. (2016). Wearable devices to support rehabilitation and social care. *Journal of Assistive Technologies*, *10*(1), 51–63. Available from https://doi.org/10.1108/jat-01-2016-0004.

Elias, A. A. (2019). Strategy development through stakeholder involvement: A New Zealand study. *Global Journal of Flexible Systems Management*, *20*, 313–322. Available from https://doi.org/10.1007/s40171-019-00217-6.

Ferguson, I., et al. (2016). Mobile health: The power of wearables, sensors, and app to transform clinical trials. *Annals of New.York Academy of Sciences, 1375*, 3–18, ISSN 0077-8923.

Kish, L. J., & Topol, E. J. (2015). Unpatients—Why patients should own their medical data. *Nature Biotechnology, 33*, 921–924.

Nanda, S. and Khattar, K. and Nanda, S. (2019). Internet of Things based remote wearable health solutions: Prospects and area of research (February 22, 2019). Proceedings of International Conference on Sustainable Computing in Science, Technology and Management (SUSCOM), Amity University Rajasthan, Jaipur - India, February 26–28. Available at SSRN: https://ssrn.com/abstract = 3351034 or http://doi.org/10.2139/ssrn.3351034.

Index

Printed in the United States
by Baker & Taylor Publisher Services